International Law,
Sustainable Development
and Water Management

International Law, Sustainable Development and Water Management

Antoinette Hildering

Eburon Publishers
2004

Title: 'International Law, Sustainable Development and Water Management'

Cover: 'Kringloop', by Antoinette Hildering
Cover design: Studio Hermkens, Amsterdam

ISBN 90 5972 055 5

Eburon Academic Publishers
P.O. Box 2867
2601 CW Delft
The Netherlands
info@eburon.nl
www.eburon.nl

Contents

List of abbreviations

CRC	Convention on the Rights of the Child
DSB	Dispute Settlement Body of the WTO
DRD	UNGA Declaration on the Right to Development
EC	European Community
ECE	UN Economic Commission for Europe
ECOSOC	UN Economic and Social Council
EIA	Environmental Impact Assessment
EU	European Union
EUWFD	EU Water Framework Directive
FAO	Food and Agriculture Organization
GATT	General Agreement on Tariffs and Trade
GEF	Global Environment Facility
ICCPR	International Covenant on Civil and Political Rights
ICESCR	International Covenant on Economic, Social and Cultural Rights
ICJ	International Court of Justice
IUCN	The World Conservation Union
ILA	International Law Association
IMF	International Monetary Fund
IWRM	Integrated Water Resources Management
WSSD	World Summit on Sustainable Development
MAI	Multilateral Agreement on Investment
NAAEC	North American Agreement on Environmental Cooperation
NAFTA	North American Free Trade Agreement
NGOs	Non-Governmental Organizations
OECD	Organization for Economic Cooperation and Development
PCA	Permanent Court of Arbitration
SADC	Southern African Development Community
UDHR	Universal Declaration of Human Rights
UN	United Nations
UN ILC	UN International Law Commission
UNCCD	UN Convention to Combat Desertification in Countries Experiencing Serious Drought and/or Desertification, Particularly in Africa
UNCED	UN Conference on Environment and Development
UNCSD	UN Commission on Sustainable Development
UNCTAD	UN Conference on Trade and Development
UNDP	UN Development Programme
UNEP	UN Environment Programme
UNFCCC	UN Framework Convention on Climate Change
UNGA	UN General Assembly
UNSC	UN Security Council

WCD	World Commission on Dams
WTO	World Trade Organization
WUAs	Water Users Associations
WWAP	World Water Assessment Programme
WWDR	World Water Development Report

Table of treaties

ECE Convention on Access to Information, Public Participation in Decision-making and Access to Justice in Environmental Matters (Aarhus Convention), Aarhus, 25 June 1998, entry into force: 30 October 2001, 38 *ILM* (1999), 517.

ECE Convention on the Protection and Use of Transboundary Watercourses and International Lakes, Helsinki, 17 March 1992, entry into force: 6 October 1996, 31 *ILM* (1992), 1312.

ECE Protocol on Water and Health to the 1992 Convention on the Protection and Use of Transboundary Watercourses and Lakes, London, 17 June 1999, not in force, 29 *EPL* (1999), 200.

General Agreement on Tariffs and Trade (GATT), Geneva, 30 October 1947, provisional entry into force: 1 January 1948, 55 *UNTS*, 308.

International Convention for the Regulation of Whaling, Washington, 1946, entry into force: 10 November 1948, 161 *UNTS*, 72.

International Convention on International Trade in Endangered Species of Wild Fauna and Flora (CITES), Washington, 3 March 1973, entry into force: 1 July 1975, 993 *UNTS*, 243, and 12 *ILM* (1973), 1085.

International Covenant on Civil and Political Rights (ICCPR), New York, 16 December 1966, entry into force: 23 March 1976, 999 *UNTS*, 171, and 6 *ILM* (1967), 368.

International Covenant on Economic, Social and Cultural Rights (ICESCR), New York, 16 December 1966, entry into force: 3 January 1976, 993 *UNTS*, 3, and 6 *ILM* (1967), 360.

Kyoto Protocol to the United Nations Framework Convention on Climate Change, Kyoto, 11 December 1997, not in force (will enter into force after the soon expected ratification by the Russian Federation), 37 *ILM* (1998), 22.

North American Free Trade Agreement (NAFTA), Washington, Ottawa and Mexico City, 17 December 1992, entry into force: 1 January 1994, 32 *ILM* (1993), 289 and 605.

Partnership Agreement between the Members of the African, Caribbean and Pacific Group of States of the one part, and the European Community and its Member States, of the other part (Treaty of Cotonou), Cotonou, 23 June 2000, entry into force: 1 April 2003, *The ACP-EU Courier*.

Protocol on Shared Watercourse Systems in the Southern African Development Community, Johannesburg 18 August 1995, entry into force: 29 September 1998, replaced by the Revised Protocol on Shared Watercourses in the Southern African Development Community, 7 August 2000, 20 *ILM* (2001), 321.

Statute of the International Court of Justice, San Francisco, 26 June 1945, entry into force: 24 October 1945, *ICJ Acts and Documents*, 4, 61.

UN Convention on the Law of the Non-Navigational Uses of International Watercourses, New York, 21 May 1997, not in force, 36 *ILM* (1997), 719.

UN Convention on the Law of the Sea, Montego Bay, 10 December 1982, entry into force: 16 November 1994, 21 *ILM* (1982), 1261.

UN Convention to Combat Desertification in Those Countries Experiencing Serious Drought and/or Desertification, Particularly in Africa (UNCCD), Paris, 17 June 1994, entry into force: 26 December 1996, 33 *ILM* (1994), 1328.

United Nations Framework Convention on Climate Change (UNFCCC), New York, 9 May 1992, entry into force: 21 March 1994, 31 *ILM* (1992), 849.

Vienna Convention for the Protection of the Ozone Layer, Vienna, 22 March 1985, entry into force: 22 September 1988, 26 *ILM* (1987), 1529.

Vienna Convention on the Law of Treaties, Vienna, 23 May 1969, entry into force: 27 January 1980, 1155 *UNTS*, 331, and 8 *ILM* (1969), 679.

Preface

> We cannot solve our problems with the same
> thinking we used when we created them.
> (attributed to Albert Einstein)

> Thus a river, viewed as a stream, is the property of the people through whose
> territory it flows, or of the ruler under whose sway that people is...
> [T]he same river, viewed as running water, has remained common property,
> so that any one may drink or draw water from it.
> (Grotius, *De juri belli ac pacis*, 1625)

> Rain does not fall on one roof alone.
> (African Proverb, Cameroon)

A growing awareness of the need to achieve sustainable development is one of the most important insights humanity has gained in recent times. The issues and dilemmas involved when considering sustainable development have always held my interest and were an important reason for my decision to study international law. In 1999, a research programme on international law and sustainable development was initiated at the Vrije Universiteit, Amsterdam. Within the research programme, climate change, fresh water, forests and investment regimes were identified as complex key areas in need of separate research. This book is the result of the research on fresh water. The dynamics of water provided me with the challenge of combining the fascinating and complex interrelationships on the ground and the need for actual responses by international law. In this book, I hope to have done justice to the research and its various aspects.

I am grateful for the support of many people, some of whom I would like to thank explicitly. First, I would like to express my gratitude to Prof. Joyeeta Gupta, Prof. Nico Schrijver and Prof. Pier Vellinga for their efforts throughout the research, guiding me in the ways of conducting research and the criteria it has to fulfil, reading through various stages of the drafts, providing time for regular meetings, and, most importantly, giving constructive criticism. I am also most grateful for the valuable comments and support of Dr. Karin Arts, Prof. Ellen Hey, Prof. André Nollkaemper, Prof. Huub Savenije and Prof. Jan Struiksma. I would also like to thank Dr. Murray Pearson for commenting my English. I would furthermore like to mention the support of my colleagues from the Faculty of Law and the Institute for Environmental Studies of the Vrije Universiteit, Amsterdam, who have not only provided feedback but also created a stimulating working environment. In this context I would especially like to mention my colleagues and friends Nienke van der Burgt, Danielle van Dam, Kenneth Manusama, Evert Neppelenbroek, Nancy Omtzigt and Mieke Tromp

Meesters. I would also like to express my appreciation to Dr. Salman Salman, who further acquainted me with the 'international water world'. I am most grateful to Dr. Ling Kan, whose several conversations were most inspiring. I would furthermore like to express my gratitude to my family, my in-laws and my friends, especially Annelies Verveen, Herman Hildering, Margreta Hildering, Lucie and Peter Oomens, and Saskia in 't Veld, for their trust in me and most certainly for their warm and understanding attitude toward me throughout these years. Most of all, I would like to thank Victor Oomens for his support, providing me with feedback on my work and thoughts, assisting me in preparing the text for publication, and creating an enabling environment.

1. Introduction

1.1 The problem

Water is a basic necessity of all life on earth. Not only is it required for the environment and its life support systems, the access to fresh water affects the very existence of human beings and their inherent dignity. In many of its functions, water cannot be substituted by alternatives, which makes it different from other natural resources such as oil.[1] Ideally, water should be available in quantities and of a quality that serves the well-being of people as well as the earth as a whole. However, according to the World Water Development Report (WWDR), 1.1 billion people do not have access to sufficient clean drinking water and 2.4 billion people lack access to adequate sanitation.[2] In addition, about half the rivers and lakes of our planet are seriously polluted, the world's wetlands are disappearing and important underground aquifers are over-exploited.[3]

Freshwater resources constitute only a small part of the world's water.[4] The gap between the supply of and demand for fresh water has been widening over the past century, due to such factors as population growth, urbanisation and an increase of water consumption per capita.[5] In addition, the amount of fresh water is reduced by paving and destruction of habitats: in both situations more water runs off to sea instead of being absorbed by the soil, fauna or flora, also possibly leading to sea-level rise.[6] Technical solutions such as desalination offer only limited solutions to the shortages considering the money and energy it involves. The shortage of fresh water and the uneven distribution of water in time and space contribute to a

---

[1] On the irreplaceable nature of water, see *e.g.*, Petrella (2001), p. 55, who argues that water is a unique source of life, comparable only with air, to which human beings need to have recourse to live. It cannot be replaced, for example, in the way that oil can substitute for coal, while the functioning of market mechanisms requires interchangeable goods and services whose comparative use value can be reflected in relative prices. See also Savenije and Van der Zaag (2002) and Savenije (2002) on the non-substitutable character of water.

[2] WWAP (2003), p. 108. This first and comprehensive World Water Development Report is the result of the joint work of 23 UN agencies and commissions dealing with water, constituting the World Water Assessment Programme (WWAP). On the numbers of people without adequate access to water see also Gleick (2000), p. 1, and UNICEF (1996), *The State of World's Children 1996*, pp. 84-85, Table 3, percent of population with access to safe water 1990-1995.

[3] Brown Weiss (1989), p. 235, elaborating on the depletion of fresh water resources, defines aquifers as: 'water-bearing strata of permeable rock, sand, or gravel.'

[4] On water and its characteristics see Chapter 2.

[5] On urbanisation, see *e.g.*, WWAP (2003), p. 14. On increase of water consumption, see Section 6.2.1.

[6] See Barlow and Clarke (2002), pp. 10-12, discussing research done by Kravčík.

number of serious global problems. Water availability varies between regions from abundance to shortage. Illustrating the uneven distribution of
available water, almost a quarter of the world's population lives in China,
where only 6 percent of the world's fresh water is located.[7] Water is especially scarce in the Middle East. In this region, people are highly dependant
on groundwater. Groundwater resources in the Arabian Peninsula – where
extraction is three times the replenishment rate – and in Israel are over-
exploited.[8] Moreover, it is estimated that Israeli settlers in the West Bank
use about four times more water than Arabs in the same area.[9] The severe
situation has made water part of political tensions as well as of peace negotiations.[10]

Water can also threaten the lives of people, for instance in the case of a
flood. According to the WWDR: 'Between 1991 and 2000 over 665,000
people died in 2,557 natural disasters, of which 90 percent were water-
related events. The vast majority of victims (97 percent) were from developing countries'.[11] Moreover, human settlements far from water resources
– for instance because of urbanisation – require transport of water during
which waste often occurs. Other factors that contribute to the water crisis
include pollution and the over-exploitation of water resources, mismanagement and failure to cooperate.

---

[7] Barlow and Clarke (2002), p. 22.

[8] See, *e.g.*, Homer-Dixon, Boutwell and Rathjens (1993), p. 22: 'Current Israeli demand
- including that of settlements in the occupied territories and the Golan Heights - is
about 2,200 mcm. The annual deficit of about 200 mcm is met by overpumping aquifers.'

[9] Postel (1996), p. 37. Israel has restricted the number of wells Arabs can drill in the
territory, the amount of water Arabs are allowed to pump and the times at which they
can draw irrigation water, see Homer-Dixon *et al.* (1993), pp. 22-23. See also Toebes
(1999), pp. 335-336: 'An example of failure to ensure access to water and sanitation
concerns the situation in the Gaza Strip and the West Bank.' On the status of Palestinian waters under international law see Abouali (1998). See also Elmusa (1995) who
argues that application of equitable apportionment would entitle Palestinians to a much
larger share than they are allowed at present.

[10] See Homer-Dixon *et al.* (1993), p. 23: 'Concerns over water access contributed to
tensions preceding the 1967 Arab-Israeli War; the war gave Israel control over most of
the Jordan Basin's water resources.' On prospects for a water regime for the Jordan
River Basin, see Baim (1997). An important example of a peace treaty that includes
water (Yarmuk and Jordan rivers) is the Treaty of Peace Between the State of Israel and
the Hashemite Kingdom of Jordan (in Article 6 and Annex II), 26 October 1994, Text
in *ILM* 34 (1995) 46. Apart from positive effects of such inclusion, Fathallah (1996), at
p. 149, rightly points out that the trend of bilateral negotiations entails the risk that
water allocation will not be regulated on a suitably larger, regional scale. See also Van
der Zaag and Vaz (2003), providing explanations why cooperation prevailed over conflict in the case of the Incomati waters shared between Mozambique, South Africa and
Swaziland. The Tripartite Interim Agreement on the Protection and Sustainable Utilisation of the Water Resources of the Incomati and Maputo Watercourses was signed by
those countries on 29 August 2002 during the WSSD.

[11] WWAP (2003), p. 12.

Where an adequate quantity and quality of fresh water is unavailable, the different uses of fresh water by different parties will entail competing interests, rights and obligations at local as well as global level.[12] The competing interests involved usually lead to trade-offs. For example, provision of drinking water and sanitation by a state for all of its population could, under certain circumstances, lead to a shortage of fresh water for agricultural purposes and thus to food-shortages. Likewise, water used for agricultural and industrial purposes may pollute drinking water supplies and the environment. For example, Lake Chad has been diminished by about 90 per cent since 1960, mainly due to irrigation.[13] In developing countries especially, the allocation of fresh water between its various uses raises problems.[14] Conflicts between meeting basic needs in the short-term and economic development in the long-term put a heavy burden on developing countries. Long-term national economic development considerations may result in large-scale development projects, which in turn may exacerbate short-term water problems of underprivileged groups.[15]

The differentiation in effects of water stress on distinctive groups of people within states is not limited to developing countries. In both developed and developing countries, problems of water can disproportionately affect particular groups of the population. Moreover, poor people are the most vulnerable to disasters like droughts and floods and to consequences of pollution and over-consumption. Not only do these situations lead to increased poverty, they also often lead to serious environmental damage. The relationship between access to water and ownership of land further makes it difficult for vulnerable groups to gain access to water. Poverty often implies lack of influence and discrimination, as in the case of the position of the Dalits in India.[16]

Such problems require responses at the community, national and international level. They have contributed to a growing awareness of the need to achieve sustainable development.[17] Sustainable development has become

---

[12] Hunter, Salzman and Zaelke (2002) at p. 769 state that: 'The resolution of competing uses over scarce fresh water supplies promises to be one of the major challenges for the next century.' That scarcity is not a necessary condition for conflict over water use can be illustrated by the Missouri River, see Tarlock (1997). On conflicts between uses of water see Chapter 2.

[13] Lake Chad is shared by Sudan, Chad, Nigeria and Cameroon.

[14] Problems of allocation of water in developing countries can be aggravated among other things by the educational situation in developing countries, or by their geographical, financial and technical position.

[15] As occurred in the 1989 conflict over the Senegal River valley between Mauritania and Senegal. See Homer-Dixon et al. (1993), pp. 19-20.

[16] Although untouchability is forbidden by law, the caste-system in practice still leads to situations in which Dalits are barred from the use of water resources so that they will not 'pollute' the water, see www.dalits.org, www.idsn.org and www.indianet.nl/Dalits.

[17] Sustainable development is defined in the scope, Section 1.3, and further elaborated upon in Chapter 3.

part and parcel of the global concern over the water crisis.[18] At the 2000 UN Millennium Summit and the 2002 World Summit on Sustainable Development in Johannesburg (WSSD), water was identified as one of the main global concerns.[19] This reflects international recognition of the global water problem that started several decades ago.[20]

As far back as in 1972, the Stockholm Conference and resulting Declaration addressed international concerns on issues that are now referred to as sustainable development.[21] The first global conference on water took place in Mar del Plata in 1977. The conference resulted in the Mar del Plata Action Plan.[22] Following, the period 1981-1990 was declared the International Drinking Water Supply and Sanitation Decade. In January 1992, a water conference took place in Dublin in preparation for Rio and resulted in the Dublin Statement on Water and Sustainable Development (Dublin Statement), including the Dublin Principles.[23] Sustainable development and freshwater resources were further discussed at global level at the 1992 United Nations Conference on Environment and Development (UNCED) at Rio de Janeiro.[24] The participating states, committed themselves to aim for sustainable development in the Rio Declaration on Environment and Development (Rio Declaration) and to implement the Agenda 21 that was also adopted at UNCED.[25] The UNCED documents reflect the

---

[18] On water and sustainable development, see Koudstaal, Rijsberman and Savenije (1992).

[19] The Millennium Summit took place in New York from 6 to 8 September 2000 as an integral part of the UN Millennium Assembly, see *UN Doc.* A/RES/53/202 (1998) and *UN Doc.* A/RES/53/239 (1999). The Millennium Declaration of 8 September 2000 was published in *UN Doc.* A/RES/55/2. For more information on the Summit see www.un.org/millennium. For the documents of the World Summit on Sustainable Development, Johannesburg, 26 August - 4 September 2002, see www.johannesburg-summit.org. The Summit is also being referred to as the Johannesburg Summit or Rio+10. The WSSD focussed on water, energy, health, agriculture and biodiversity (WEHAB). On water issues at the WSSD, see *e.g.*, HRH The Prince of Orange (2002).

[20] Milestones on the road to sustainable management of water resources during conferences and in decisions can be found in Chapter 2 of WWAP (2003).

[21] Report of the UN Conference on the Human Environment (1972), *UN Doc.* A/CONF/48/14/rev. 1.

[22] *Report of the United Nations Water Conference, Mar del Plata*, 14-25 March 1977, United Nations Publications: New York, E/77/II/A/12.

[23] The Dublin Principles refer to: the finite, vulnerable and essential nature of water; the importance of participatory water development and management; the central position of women; and the economic value of water in all its competing uses. The Dublin Statement is available through www.wmo.ch/web/homs/documents/english/icwedece.html.

[24] Report of the United Nations Conference on Environment and Development, Rio de Janeiro, 3-14 June 1992, A/CONF.151/26.

[25] Rio Declaration on Environment and Development, Annex I to the Report of the United Nations Conference on Environment and Development, Rio de Janeiro, 3-14 June 1992, *UN Doc.* A/CONF.151/26 (Vol. I). Agenda 21, *UN Doc.* A/CONF.151/26 (Vol. I, II and III), www.un.org/esa/sustdev/documents/agenda21/english/agenda21toc.

complexity of sustainable development of freshwater resources by underlining the need to balance many interests. Chapter 18 of Agenda 21, on the protection of the quality and supply of freshwater resources, presents a policy framework on freshwater resources. It sets out the aim of ensuring that:[26]

> adequate supplies of water of good quality are maintained for the entire population of this planet, while preserving the hydrological, biological and chemical functions of ecosystems, adapting human activities within the capacity limits of nature and combating vectors of water-related diseases.

In March 2000, the Second World Water Forum took place in The Hague, positioning the water crisis high upon the international agenda.[27] The importance of water for sustainable development was reaffirmed in the Millennium Summit, which resulted in the Millennium Declaration, including the International Development Target:[28]

> To halve, by the year 2015, the proportion of the world's people whose income is less than one dollar a day and the proportion of people who suffer from hunger and, by the same date, to halve the proportion of people who are unable to reach or to afford safe drinking water.

The commitment in the Declaration to adopt 'a new ethic of conservation and stewardship', includes: 'To stop the unsustainable exploitation of water resources by developing water management strategies at the regional, national and local levels, which promote both equitable access and adequate supplies.'[29] The importance of water for sustainable development was further emphasised at the 2001 international freshwater conference

---

[26] Agenda 21, Chapter 18.2.

[27] See www.worldwaterforum.org. The World Water Forums are not UN meetings, but a gathering of experts on water and other interested parties, including ministers, scientists, indigenous people, NGOs, intergovernmental organizations, corporations and journalists. The First World Water Forum took place in 1997 at Marakech. At the Second World Water Forum, over 5000 people participated. The Second and Third World Water Forums included a Water Fair and a parallel Ministerial Conference. The world water fora were initiated by the World Water Council (WWC), see www.worldwatercouncil.org. The fora are also linked to the Global Water Partnership (GWP) and the World Commission on Water (WCW) for the 21st century. While the GWP, WWC and WCW are ostensibly objective, Barlow and Clarke (2002), pp. 157-158, argue that these bodies essentially represent corporate and neo liberal interests.

[28] UN Millenium Declaration Chapter III, para. 19, on development and poverty eradication, see note 19.

[29] UN Millenium Declaration Chapter VI, para. 23, on protecting our common environment.

held in Bonn in preparation of the WSSD.[30] One of the key commitments made at the WSSD was to halve, by the year 2015, both the number of people without access to safe drinking water (affirmation of the Millennium Development Goal) and the number of people who do not have access to basic sanitation. Furthermore expressing the global importance of water, the year 2003 was proclaimed as the International Year of Freshwater by the UN General Assembly (UNGA).[31] In 2003, the Third World Water Forum took place at Kyoto, providing a forum for states and non-state actors.[32]

Despite all these declarations and policies, the problems relating to water remain unsolved: human health is suffering from water-related diseases, while the demand for water per capita still increases, and the degradation of ecosystems continues. As pointed out by the WWDR, the water crisis is mainly a crisis of governance, due to inadequate leadership and insufficient awareness of the magnitude of the problems with the world's population.[33]

1.2   Water and international law: the research questions

The problems related to fresh water have an impact on people as well as on the environment. Public international law (hereafter international law) is one instrument that may help to bring about a balance between the different interests in play in such a way as to protect both people and the environment.[34] The potential relevance of international law to water management is stressed by the existence of over 260 international river basins, also referred to as watersheds or catchment areas, shared by two or more states.[35] Moreover, international underground waters can be found

---

[30] See www.water-2001.de.

[31] *UN Doc.* A/RES/55/196, International Year of Freshwater, 2003.

[32] See note 27.

[33] See *e.g.*, WWAP (2003), p. 383: 'The water crisis is essentially about how we as a society and as individuals perceive and govern water resources and services.', p. 528 WWAP, and the WWDR Executive Summary, online on unesdoc.unesco.org. See on global sustainable development governance Gupta (2002) and Gupta and Hisschemöller (1997).

[34] See for general principles of international law, Shaw (1997), Schachter (1991) and Brownlie (1998).

[35] It is estimated that 263 river basins are shared by two or more states in which about 40 percent of the population lives, see WWAP (2003), p. 10. At p. xix, WWAP speaks of over 300 rivers that cross national boundaries. Since a river basin can include more than one river, these numbers can co-exist. Gleick (2000) lists 261 international river basins in Table 7, p. 219-238, referring to the analysis of Wolf, Natharius, Danielson, Ward and Pender (1999). According to Hunter *et al.* (2002), p. 769, worldwide more than 200 river basins are shared by two or more states. Altogether, international river basins comprise almost half the land area of the world, Antarctica excluded, see Gleick (2000), p. 219. See Section 6.4.1 of this book on the catchment area approach.

throughout all continents, such as the Nubian aquifer underlying Chad, Egypt, Libya and Sudan.

Contemporary international law can be traced back to the emergence of the modern nation-state, after the Peace of Westphalia in 1648. In essence, it is state-oriented and based on features such as sovereignty and responsibility of states and equality of and reciprocity between states.[36] Since there exists no supranational government and states are sovereign, in principle international law is created by the explicit or implicit consent of states. This basic structure of international law – lacking a division of powers comparable to the *trias politica* – explains the limitations in possibilities for enforcement of international law. However, during the last few decades many of its features have undergone changes. International law is no longer limited to states, but also addresses international organizations and non-state actors. Moreover, sovereignty of states increasingly refers not only to rights but to duties as well.[37]

Extensive research has been undertaken on the (emerging) international law on sustainable development and international water law separately.[38] The interaction between international water law and international law on sustainable development has received much less attention. The international law on freshwater resources does not necessarily contribute to sustainable development and although international law on sustainable development is emerging, its application to freshwater resources remains unclear. Moreover, the uneven distribution of water at the national and local level and its effects on vulnerable population groups raises questions on the right of access to water for all people.[39]

Transparency on the applicable principles is lacking due to the many fields of international law and other disciplines involved. The principles of international law that stimulate sustainable development of freshwater resources need to be further identified and analysed. Most of all, further integration of the relevant principles of international law into a transparent framework is needed. The various ingredients need to balance the social, economic and ecological pillars of sustainable development. Providing a comprehensive legal framework within which such a balance is to be found, is the challenge undertaken in this study. Such a legal framework

---

[36] See *e.g.*, Schrijver (2000).

[37] On duties that arise from the principle of permanent sovereignty, see Schrijver (1997), pp. 391-392.

[38] In the field of international law on sustainable development see ILA Committee on Legal Aspects of Sustainable Development (2002) and the First Report of the ILA Committee on International Law on Sustainable Development (2004), www.ila-hq.org. In the field of international water law see the Fourth Report of the ILA Committee on Water Resources Law (2004), McCaffrey (2001), Tanzi and Arcari (2001), Brans, De Haan, Nollkaemper and Rinzema (1997), Bruhács (1993) and Caponera (1992). Both fields of international law are introduced in Chapter 3 of this study.

[39] See Section 4.3.2 of this study on social groups. See on access to water as a human right Section 4.2.

can guide policy-makers and people in general to contribute to freshwater management so as to achieve sustainable development.

The general research question of this study is: In what way can international law contribute to the achievement of sustainable development in and through water management? More specifically, this study addresses the following three research questions:

- *Which principles of international law can be instrumental in achieving sustainable development in water management?;*
- *How do they relate to one another?; and*
- *How can they jointly contribute to a more sustainable development of freshwater resources?*

1.3   The scope

This study focuses primarily on freshwater resources. 'Freshwater resources are an essential component of the Earth's hydrosphere and an indispensable part of all terrestrial ecosystems.'[40] The reason for limiting this study to this part of the water cycle is twofold. Firstly, although freshwater resources are part of the same hydrological cycle as the oceans and water in the atmosphere, the natural characteristics of freshwater resources, and therefore their uses, differ. Secondly, the law of the sea is relatively well developed in comparison with the fragmented international law on fresh water, and is a well articulated field of international law. The need for compatible regimes governing the seas and fresh water, as well as other interrelated areas, is nevertheless an essential one. This need can be illustrated by, for example, the pollution of the oceans by land-based sources such as through rivers.

Hereafter, water refers to fresh water unless stated otherwise. International law on fresh water will also be referred to as international water law. Unless otherwise stated, when discussing water resources reference is made to both surface and groundwater, considering their interrelationship and the need for an integrated approach.

Water and the relevant international law are analysed in relation to the goal of sustainable development.[41] Sustainable development in the field of water management is not to be confused with sustainable use of water resources.[42] Sustainable utilization is important but only one aspect of sustainable development: a lasting use of water resources does not require the

---

[40] Chapter 18 of Agenda 21, para. 18.1.
[41] On sustainable development as an aim see Section 3.1 of this study.
[42] On sustainable use of water resources, see *e.g.*, Hey (1995).

existence of people, let alone necessarily involve developmental elements such as equity or an adequate standard of living for all.[43]

Whether or not and to what extent the goal of sustainable development is expressed in international law and implemented in practice first and foremost depends on the willingness of the international community. The reiterated commitment of policymakers and practice by states and other actors lead us to the first and foremost axiom of this research: that the international community seriously wants to achieve and consolidate sustainable development, is aware of the necessity to do so, and is committed to its realisation.

The legal dimension of sustainable development expressed by the international community was not very clearly articulated in the nineties. Principle 27 of the 1992 Rio Declaration calls for the further development of international law in this field: 'States and people shall cooperate in good faith and in a spirit of partnership in the fulfilment of the principles embodied in this Declaration and in the further development of international law in the field of sustainable development'.[44]

The standard definition of sustainable development is the Brundtland definition: 'Sustainable development is development that meets the needs of the present without compromising the ability of future generations to meet their own needs.'[45] Taking into account the more recent insights into sustainable development, including an increased acknowledgment of both developmental and environmental concerns and human rights, the International Law Association (ILA) adopted the New Delhi Declaration of Principles of International Law Relating to Sustainable Development (ILA New Delhi Declaration). This Declaration includes a definition of sustainable development that will be the starting point of this research. The Declaration expresses the view that:[46]

---

[43] Nevertheless, definitions of sustainable use often do include such elements and practically overlap with sustainable development, as does Article 3.19 of the 2004 ILA Berlin Rules, defining sustainable use as 'the integrated management of resources to assure efficient use of and equitable access to waters for the benefit of current and future generations while preserving renewable resources and maintaining non-renewable resources to the maximum extent reasonably possible.' See Section 3.3.1 of this study on the Berlin Rules.

[44] See also Chapter 39 of Agenda 21 on international legal instruments and mechanisms.

[45] World Commission on Environment and Development (1987), p. 43. The definition is clearly based on an earlier reference to sustainable benefit to present generations and needs of future generations in IUCN-UNEP-WWF (1980), *World Conservation Strategy: Living resource conservation for sustainable development*, Geneva/Nairobi. This report also formed the basis of the World Charter for Nature, adopted by the UNGA on 28 October 1982, A/RES/37/7, which in its preamble emphasises that natural resources are to be used in a way 'which ensures the preservation of species and ecosystems for the benefit of present and future generations'.

[46] ILA New Delhi Declaration of Principles of International Law Relating to Sustainable Development, Annex to the letter dated 6 August 2002 from the Permanent Represen-

the objective of sustainable development involves a comprehensive and integrated approach to economic, social and political processes, which aims at the sustainable use of natural resources of the Earth and the protection of the environment on which nature and human life as well as social and economic development depend and which seeks to realize the right of all human beings to an adequate living standard on the basis of their active, free and meaningful participation in development and in the fair distribution of benefits resulting therefrom, with due regard to the needs and interests of future generations...

The challenge of achieving sustainable development focuses on human beings and their behaviour. A balance needs to be reached between nature and people and also among people. Sustainable development as defined above cannot be realised when an adequate standard of living is denied to the larger part of the world's population. According to the Brundtland Commission, 'the essential needs of the world's poor' are to be given overriding priority.[47] Moreover, as voiced by Sachs: 'Protecting the rights of the most vulnerable members of our society, in other words, is perhaps the best way we have of protecting the right of future generations to inherit a planet that is still worth inhabiting.'[48] Therefore, combined with the fact that water problems have a particular impact on vulnerable groups of people and developing countries, special regard is given to those people and countries.

Furthermore, this research takes a pluralistic approach to international law. The sustainable development of water resources cannot be achieved by states alone. The interests and roles of non-state actors also have to be taken into account. This does not imply a denial of the sovereignty of states over water resources, but rather recognizes the importance of public participation and awareness in the process of water management. This implies that the research focuses on the international legal aspects, including the relationships between various administrative levels of governance and evaluates the meaning of applicable international law principles at these levels. Moreover, for international law to be of assistance it has to be viewed within its broader context including other disciplines of specific

---

tative of Bangladesh to the UN and the Chargé d'affaires a.i. of the Permanent Mission of the Netherlands to the UN addressed to the Secretary-General, *UN Doc.* A/57/329, p. 3. The Declaration can be found on www.un.org/ga/57/document.htm. The International Law Association (ILA) adopted the Declaration by consensus by Resolution 2003/3 on 6 April 2002. The Declaration was prepared by the ILA Committee on Legal Aspects of Sustainable Development (Chairman: Dr Kamal Hossain, Bangladesh; Rapporteur: Prof Nico Schrijver, The Netherlands), and can be seen as a follow-up to the 1986 ILA Seoul Declaration on the Progressive Development of Principles of Public International Law relating to a New International Economic Order.

[47] World Commission on Environment and Development (1987), p. 43.

[48] Sachs (1995), p. 55.

importance to the subject, such as hydrology, sociology, economics, biology, and politics.

Exploring the notion of sustainable development and the relevant international law and relating it to water management calls for a thorough analysis of a whole range of issues involved. This study examines a very broad range of subjects and how they relate rather than presenting an in-depth research of the more specific legal, social, political, hydrologic, eco-logic or economic areas or of case studies. Where necessary, this is supplemented by references to other studies providing more detailed information and insights.

1.4   The methodology

The methodology adopted in this study can be divided into four steps:

The first step is the identification of the present state of the concept of sustainable development in international water law.

The status of concepts and principles in international law depends on the sources of international law.[49] The classic sources of international law are reflected in Article 38 of the Statute of the International Court of Justice (ICJ), which refers to: international conventions, whether general or particular, establishing rules expressly recognised by states (treaty law); international custom, as evidence of a general practice accepted as law (customary international law); the general principles of law; and as subsidiary means for the determination of rules of law (secondary sources of international law), judicial decisions and the teachings of the most highly qualified publicists of the various nations (doctrine).[50] Emerging but not yet well-established law can be referred to as *de lege ferenda*, opposite to *de lege lata* that encompasses well-established applicable law.

Over the last decades, there has been an enormous increase in bilateral and multilateral agreements between states. The main rules on treaty law are laid down in the Vienna Convention on the Law of Treaties.[51] Article 2 of this Convention defines a treaty as 'an international agreement con-cluded between States in written form and governed by international law, whether embodied in a single instrument or in two or more related instru-ments and whatever its particular designation.' Section 3 presents the rules of interpretation of a treaty, referring in Article 31 to the ordinary meaning

---

[49] See Shaw (1997), Chapter 3 and Brownlie (1998), Chapter 1.

[50] Statute of the International Court of Justice, San Francisco, 26 June 1945, entry into force: 24 October 1945, *ICJ Acts and Documents*, No. 4, 61. Under Article 93 of the UN Charter, all UN members are parties to the Statute of the ICJ (as on 4 October 2004: 191). See www.icj-cij.org.

[51] Vienna Convention on the Law of Treaties, Vienna, 23 May 1969, entry into force: 27 January 1980, 1155 *UNTS*, 331, and 8 *ILM* (1969), 679.

of its terms and to its context, in addition to the text, including its preamble and annexes, such as certain related agreements. Additional means of interpretation are included in Article 32, which includes the preparatory work of the treaty. According to Article 18 of the Vienna Convention on the Law of Treaties, signatories must not act against the object and purpose of a treaty even if it is not in force or not yet ratified by the signatory.

In the case of codification of customary international law in a treaty, the rules involved also remain binding under customary international law. This also implies that the rules involved are binding the parties to the treaty whether or not the treaty is in force. Likewise, states not party to the treaty are bound by customary international law. Customary international law is established by state practice (*usus*), together with the conviction that the principle presents a legal obligation (*opinio juris sive necessitatis*).[52]

Especially in the absence of treaties or customary international law, general principles of law become all the more important. General principles of law are harder to define in detail, but in broad terms are those principles perceived by states as underlying concepts of conduct that guide the legal system. General principles of law can refer to both a general principle of law appearing in municipal systems or a general principle of international law.[53] General principles of law are rare.[54] Examples of such principles include the principles of justice, equity, good-neighbourliness, and the prohibition of abuse of rights.

The subsidiary sources complement the aforementioned ones. The judgments or advisory opinions of the ICJ, as well as judgments of other international tribunals and national courts are only binding between parties but they can be of great significance in the interpretation and further definition of international law. Moreover, the separate opinions to the ICJ judgments not necessarily formulate but do provide a prominent voice as to the state of law. The writings of authors of great authority also influence the interpretation and evolution of international law, for example, by identifying customary law. The work of institutions such as the Institut de Droit International and the ILA has also enriched and assisted the progressive development of international law.[55]

Additional sources of international law have come into existence since the formulation of Article 38 of the Statute of the ICJ. Whether or not these sources present separate sources or are covered by the sources of Article 38, as evidence of customary law or based upon treaty law, is subject of debate. Such sources include the resolutions of the Security Council

---

[52] See Shaw (1997), pp. 56-73.

[53] *Cf.* Shaw (1997), pp. 78-79.

[54] *Cf.* Shaw (1997), pp. 77-78, on general principles of law, who states at p. 78: 'most writers are prepared to accept that the general principles do constitute a separate source of law but of fairly limited scope, and this is reflected in the decisions of the Permanent Court of International Justice and the International Court of Justice.'

[55] See respectively www.idi-iil.org and www.ila-hq.org.

(UNSC), which can be binding upon states. Moreover, international organizations are attributed implicit powers that go beyond those explicitly consented to by states.[56] These implied powers were formulated in the ICJ Judgment in the *Reparation for Injuries* case, in which case the ICJ found that the UN had obtained powers implicitly conferred on it by the parties: 'Under international law, the Organization must be deemed to have those powers which, though not expressly provided in the Charter, are conferred upon it by necessary implication as being essential to the performance of its duties'.[57]

Furthermore, instruments of "soft law", although not legally binding *stricto senso*, can have enormous impact. According to Birnie and Boyle, soft law records, in written form, norms agreed to by states or international organizations that leave a considerable degree of discretion in interpretation and implementation. Soft law can be an additional secondary source of law, filling the gaps, guiding interpretations, or revealing the emergence of international law.[58] Such sources include resolutions of the UNGA and documents resulting from world summits.[59] The ICJ in the *Nicaragua* case and the *Advisory Opinion on the Legality of Nuclear Weapons* underlined the importance of sources of international law such as UNGA Resolutions. Moreover, documents resulting from world summits, such as the Rio Principles, can carry considerable normative weight and/or embody a programme of action for the international community.[60] Their impact can be witnessed by the many actions undertaken at various administrative levels to adapt policies as well as law in response to the call for sustainable development.[61] Furthermore, in the light of the development of community interests, instruments other than those explicitly or implicitly consented to by states are becoming of increased importance.[62]

The aforementioned identification of the present state of the concept of sustainable development in international water law includes an analysis of: treaties (such as the UN Watercourses Convention and ECE conventions); customary international law (as established in ICJ judgments and

---

[56] This so-called 'implied powers' theory goes beyond the classical version of state consent. See Schermers and Blokker (2003), Chapter 3.

[57] *Reparations for injuries suffered in the service of the United Nations* case, Advisory Opinion of 11 April 1949, ICJ Reports 1949, p. 182.

[58] Birnie and Boyle (2002), pp. 24-27, refer at p. 25 to half-way stages in the law-making process, including resolutions, declarations of principles, recommendations and guidelines, often within the context of "framework" treaties, furthermore stating on "soft law" instruments that: 'they may provide good evidence of *opinio iuris*, or constitute authoritative guidance on the interpretation or application of a treaty, or serve as agreed standards for the implementation of more general treaty provisions or rules of customary law.'

[59] On the status of UNGA resolutions and declarations, see Birnie and Boyle (2002), pp. 22-24.

[60] Haas (2002).

[61] See Shelton (2000) on the role of non-binding norms.

[62] Section 3.4.3 of this study will elaborate on community interests.

addressed in literature); resolutions and documents of UN bodies (such as UNGA resolutions and UN ILC documents); law-making activities and policy documents of regional organisations (such as EU directives); declarations and resolutions resulting from conferences (*e.g.* documents that resulted from UNCED and the World Water Forums); and activities and documents of NGOs (*e.g.* work of the ILA and IUCN). It moreover includes a survey and analysis of related literature and an interdisciplinary study of sustainable development.

The second step is the development of a comprehensive framework of principles of international law. This framework can, for example, be used for analysis of legal agreements to see whether or not they support the notion of sustainable development. This framework is the result of a process of structuring and analysing available information. In this process, principles and problems are classified into categories and policy-levels.

To enable analysis, the sustainable development and water management principles and problems are classified into three categories: social, economic and ecological. The choice to use these categories is mainly based upon their correspondence to the three so-called "pillars" of sustainable development. Moreover, other possible categories were found less suitable for the research. The "Triple P" concept of people, planet and profit is often used within the context of corporate social responsibility.[63] Broadly speaking, 'people, planet and profit' refers to the social, ecological and economic elements of sustainable development. However, the concept can be confused with other concepts also referred to as "PPP", such as public-private partnerships and the polluter pays principle.[64] Another option for categorising the elements of sustainable development is into: development, environment and human rights. The original binary split into a developmental and an environmental component has during the mid-nineties become supplemented by the dimension of human rights, resulting in a division comparable to that of social, economic and ecological aspects. However, social and economic interests remain intertwined in the developmental component, complicating analysis. In addition, 'ecological' can be argued to be preferred over 'environmental' since it more clearly refers to the natural environment instead of overall surroundings.

The sequence of dealing with the pillars is not intended to reflect a hierarchy. On the contrary, as "pillars" already suggests, all pillars are placed next to each other since they are of equal importance. The order of discussing the pillars in this study – social, economic and ecological – can be

---

[63] See www.triple-p.org and on corporate social responsibility, www.irene-network.nl and www.unglobalcompact.org.
[64] The results of a search on the internet on 3P include a 'pollution prevention pays program' and on TripleP include a 'positive parenting program' and a 'planning, programming and performance' assessment. In World Bank (2003), p. 28, reference is also made to 'People, Planet, and Prosperity'.

explained by the time dimension involved: social needs include the direct urgent need for water by people; economic interests often relate to mid-term interests in particular; while many ecological effects only emerge in the longer-term. This time differential – together with the central position usually granted to humans – is frequently reflected in policies, prioritising social needs, followed by economic needs, and finally ecological needs. However, all pillars can be designed for the short-, mid-, as well as the long-term. And in the final analysis, of course, the global ecosystems support all life and activities. A bias against one of the pillars of sustainable development is unlikely to lead to long-term effectiveness for present and future generations.

The division into a social, economic and ecological category is made for analytical reasons, and the pillars or categories are to be combined in order to achieve sustainable development. As stated in the WSSD Plan of Implementation: 'These efforts will also promote the integration of the three components of sustainable development – economic development, social development and environmental protection – as interdependent and mutually reinforcing pillars.'[65]

The impact of sustainable development and water management principles and problems at the various policy-levels is examined throughout the research. Policy-making is divided into community, national, and international levels in order to highlight the different demands made by each level to international law as well as to show the impact of international law principles at the various policy-levels. In line with international law, "international" includes regional and interregional, involving two or more states unless otherwise stated. In terms of hydrology, usually the local or "community" level stands for a small watershed, "national" for watershed, and "international" for catchment basin up to the global hydrological cycle. When referred to in this book, the principle of subsidiarity means that only those tasks are performed at a certain policy-level which cannot be performed at a more local level.[66]

---

[65] WSSD Plan of Implementation, X. Institutional framework for sustainable development, para. 2, where it continues: 'Poverty eradication, changing unsustainable patterns of production and consumption, and protecting and managing the natural resource base of economic and social development are overarching objectives of, and essential requirements for, sustainable development.' The Plan of Implementation is included in the Report of the World Summit on Sustainable Development, *UN Doc.* A/CONF.199/20.

[66] Within the European Union the principle of subsidiarity, in relation to the Community and its member states, is defined in Article 3b of the Treaty establishing the European Community:

'In areas which do not fall within its exclusive competence, the Community shall take action, in accordance with the principle of subsidiarity, only if and in so far as the objectives of the proposed action cannot be sufficiently achieved by the member states and can therefore, by reason of the scale or effects of the proposed action, be better achieved by the Community'.

As a third step, this book has been constructed on the basis of the new structure. The methodology and the developed framework are therefore reflected in its structure.

The problems resulting from the competing uses of fresh water are analysed and response options are explored in Chapter 2. Next, a description and evaluation of international law on sustainable development and international water law are undertaken in Chapter 3. The principles of international law are categorised by the separate pillars of sustainable development (Part II).

Principles of international law that mainly relate to the concept *access to water* and focus on the distribution of water, are classified under the social category in Chapter 4. Distribution of water entails basic water needs of people, the correlation between water availability and the eradication of poverty and equitable allocation within and between generations. Principles of international law categorised within the social pillar of sustainable development are a (human) right to water at the community level, eradication of poverty at the national level, and the principle of equity at the international level.

Principles of international law that mainly relate to the concept *control over water* are classified under the economic category in Chapter 5. Principles of international law categorised within the economic pillar of sustainable development are a right to use water at the community level, water as an economic good at the national level, and that of a supportive and open international economic system at the international level.

Principles of international law that mainly relate to the concept *protection of water* and focus on sustainable management of water resources, are classified under the ecological category in Chapter 6. Principles of international law categorised within the ecological pillar of sustainable development are a duty to protect water at the community level, protection of the environment at the national level, and ecological integrity at the international level.

The development of the framework requires the combination of the three pillars of sustainable development, characterised by principles of international law and key concepts bridging those pillars (Part III). In Chapter 7 the social and economic elements are combined into the concept *development through water*. Development through water includes the right to development at the community level, the right of self-determination at the national level, and the principle of common but differentiated responsibilities at the international level.

The social and ecological elements are combined into the concept *life support by water*. Life support by water includes the right to a healthy environment at the community level, the precautionary principle at the national level, and the principle of eco-justice at the international level.

The combination of economic and ecological elements has resulted in the concept *sustainable use of water*. Sustainable use of water calls for the application of the polluter and user pays principle at the community level, the no-harm principle at the national level, and the principle of common heritage or concern of humankind at the international level.

In Chapter 8, the combination of all foregoing principles leads to the concept *sustainable development of water*, including human rights and duties at the community level, qualified sovereignty of states at the national level, and a modified principle of equitable and reasonable utilization of water at the international level. The integrated framework is referred to as *guardianship over water*.

The fourth step, an overall assessment of how international law can contribute to the achievement of sustainable development in and through water management, is undertaken in the concluding Chapter that contains the summary and conclusions of this study. It results in three key conclusions and three key recommendations on international law for sustainable development in water management.

1.5   Book outline

Part I introduces and defines the terms of reference in addressing the uses of freshwater resources (Chapter 2), and analyses the way in which sustainable development is embedded in international law (Chapter 3).

In Part II, an analysis is undertaken within each of the sustainable development pillars to identify the demands on international law made by water as respectively a social (Chapter 4), economic (Chapter 5), and ecological (Chapter 6) good. Chapter 4 addresses the protection under international law of basic human needs, required when regarding water as a social good. The way international law can contribute to effective and efficient use of water (water as an economic good) without compromising basic human needs is identified in Chapter 5. Chapter 6 discusses ways in which international law can protect the ecosystems in order to balance the social and economic requirements.

The combination of principles of international law relevant to sustainable development of freshwater resources is the subject of Part III. Chapter 7 bridges the gap between the social and economic, the social and ecological and the economic and ecological pillars of sustainable development. In Chapter 8 all three pillars of sustainable development are combined. The principles are presented in a new and comprehensive framework, operationalised through a Draft Declaration addressed to all concerned, a pricing mechanism and a possible application to legal instruments. The framework of Chapter 8 provides the main answer to the research questions. The concluding Chapter contains the summary and conclusions of this study.

The relevance of this study to international law can partly be attributed to the development of a new and comprehensive legal framework. This framework reveals which principles of international law can assist in achieving sustainable development and clarifies their relationship with each other. The framework combines the provision of transparency – offering increased legal guidance – with an integrated approach – taking into account the numerous aspects of water and sustainable development. The framework could, for example, be used by lawyers and policy-makers to assess the adequacy of international law instruments and of water management in achieving sustainable development.

The appreciation of all pillars of sustainable development, the analysis of classic and emerging principles of international law and the inclusion of the geographic reality of differences in policies at the community, national and international levels throughout the study further contribute to the relevance of this book.

PART I. SETTING THE SCENE

2. Uses of freshwater resources

2.1 Water

Practically all water is part of the hydrological cycle.[1] The oceans contain about 96.5% of the world's water and only 2.5% is estimated to be fresh water. Freshwater resources are those resources that contain water with such a low level of salt that they are suitable for uses such as drinking. Fresh water and salt water can mix, such as in coastal areas, resulting in brackish water. Fresh water can be categorised as brown water (groundwater), green water (such as in plants), and blue water (surface water in, for example, rivers and rain). Almost 70 percent of the fresh water is located in ice sheets and glaciers and nearly 30 percent is groundwater.[2] Only about 0.3% of the fresh water is available for uses by humankind, of which the largest part consists of groundwater and a small part can be found in rivers and lakes.[3] According to Bertels, Aiking and Vellinga, global human withdrawal and use of fresh water amounts to 3,240 km[3] each year and over the past 30 years has increased more than 35-fold.[4]

This Chapter identifies the uses of freshwater resources, classifies them and discusses the potential for conflict of interests. For the sake of analysis as elaborated already in Chapter 1, the uses of water under discussion are classified in a social, economic and ecological category. The main balance of interests will take place between those categories. However, the categories are not mutually exclusive and the uses within various categories can be mutually supportive. Moreover, there may be conflicts between uses of freshwater resources within a single category.

2.2 Social uses

Uses categorised under the social pillar of sustainable development are those that serve people's basic needs, domestic uses and food production, and those that serve cultural purposes, as a form of social interaction.

---

[1] See Section 2.4.2. of this study for a fuller treatment of the hydrological cycle.
[2] WWAP (2003), p. 67. See p. 68, Table 4.1, for the distribution of water across the world.
[3] See Gleick, Burns, Chalecki, Cohen, Cushing, Mann, Reyes, Wolff and Wong (2002), pp. 237-242, Table 1, on Total Renewable Freshwater Supply, by Country.
[4] Bertels, Aiking and Vellinga (1999), pp. 129-130.

2.2.1   Domestic uses

Domestic uses include drinking water and sanitation.[5] Domestic uses, especially drinking water, require a specifically high standard of water quality.[6] Drinking water is defined in the ECE Protocol on Water and Health to the 1992 Convention on the Protection and Use of Transboundary Watercourses and International Lakes (ECE Protocol on Water and Health) as 'water which is used, or intended to be available for use, by humans for drinking, cooking, food preparation, personal hygiene or similar purposes'.[7] The ECE Protocol also defines sanitation as 'the collection, transport, treatment and disposal or reuse of human excreta or domestic waste water, whether through collective systems or by installations serving a single household or undertaking'.

Availability of water for domestic uses touches upon the existence of human beings and their inherent dignity. The basic need for water is estimated as 40-50 litres per day per person, the exact amount further depending on factors such as climate.[8] Gleick recommends an average of 50 litres per person per day to meet human domestic needs: 5 litres for drinking water, 20 litres for sanitation services, 15 litres for bathing, and 10 litres for food preparation.[9] As stated in Chapter 1, there are 1.1 billion people that do not have access to sufficient potable water and 2.4 billion people have no access to adequate sanitation. The lack of access to clean drinking water and adequate sanitation results in major health problems. The incidence of diseases is increased by exposure to polluted water, especially in developing countries where half the population is said to be exposed to

---

[5] See on sanitation in cities, *e.g.*, United Nations Human Settlement Programme (2003).

[6] See Gleick, Burns *et al.* (2002), pp. 273-275, Table 6, on Reported Cases of Dracunculiasis by Country, 1972 to 2000, pp. 276-277, Table 7, on Reported Dracunculiasis Cases, Eradication Progress, 2000 and pp. 278-279, Table 8, on National Standards for Arsenic in Drinking Water.

[7] The ECE Protocol on Water and Health to the 1992 Convention on the Protection and Use of Transboundary Watercourses and International Lakes, London, 17 June 1999, 29 *EPL* (1999), 200, was adopted on the occasion of the Third Ministerial Conference on Environment and Health held at London and is not yet in force. Status as on 4 October 2004: 13 parties and 36 signatories.

[8] The Academic Council on the United Nations System (ACUNS) in its draft plan of action on water, energy, poverty alleviation, and governance – *Plan of Action for Johannesburg: The Development-Environment Nexus*, distributed at PrepCom III that took place at UN Headquarters on 28 March 2002 – suggests 40 litres of water per person per day as a minimum daily requirement, to be made available within 50 meters from a household. Excluding cooking, bathing and basic cleaning, the WHO, World Bank and US Agency for International Development recommended 20 to 40 liters per person per day to be located nearby, see Smets (2000), p. 249.

[9] Gleick (2000), p. 11, Table 1.1. See also Gleick, Burns *et al.* (2002): pp. 252-260, Table 3, on Access to Safe Drinking Water by Country, 1970 to 2000; pp. 261-269, Table 4, on Access to Sanitation by Country, 1970-2000; and pp. 270-272, Table 5, on Access to Water Supply and Sanitation by Region, 1990 and 2000.

such pollution.[10] In developing countries about 80 percent of diseases are water related.[11] It is estimated that every day 14 to 30 thousand people die from diseases related to water.[12] Children are especially vulnerable to water-related diseases: in 1998, of the 2.2 million people that died from diarrhoeal diseases, 1.8 million were children under the age of five.[13]

Providing drinking water and sanitation to all people requires enormous efforts and finances, but not more than Europeans are said to spend annually on ice cream.[14] Combating pollution and increasing hygiene are other important instruments to reduce the number of water-related diseases. For example, hand washing with soap removes faeces and could possibly result in more than a 45% reduction in the incidence of diarrhoea.

2.2.2   Food production

Food production refers to agricultural uses, cattle-breeding, fishery and fish-breeding. Food production requires large quantities of fresh water. A ton of harvested grain takes about 1,000 tons of water.[15] Agricultural uses account for about 65 to 70 percent of human freshwater use. The quality of water for agriculture can be less than for domestic uses, offering options of using recycled water. Sixteen percent of the croplands worldwide that are irrigated is estimated to account for 40 percent of food production.[16] According to the WWDR, '[t]he area of irrigated land more than doubled in the twentieth century.'[17] Despite the rapid increase of food production accompanying population growth, at present nearly 800 million people are undernourished.[18] Not only is agriculture the main user of water, it is said to be the main polluter of water as well.

Application of methods such as drip irrigation, less meat consumption and maybe even salt water agriculture could reduce freshwater use. Agriculture is nevertheless expected to remain the largest user of water. Both food security (enough supply) and food safety (quality of food) are at risk because of the water crisis, aggravated by, *e.g.*, urbanisation. For example,

---

[10] WWAP (2003), p. 11.

[11] See WHO/UNICEF Joint Monitoring Programme for Water Supply and Sanitation (JMP), Global Water Supply and Sanitation Assessment 2000 Report, at www.who.int/water_sanitation_health/Globassessment/ GlobalTOC.htm.

[12] Gleick (2000), p. 1. See also Hunter, Salzman and Zaelke (2002), p. 826: 'Water-borne diseases continue to be among the leading causes of death in many developing countries.'

[13] WWAP (2003), p. 36.

[14] About US$ 11 billion, supposedly US$ 2 billion more than needed for drinking water and sanitation. See, *e.g.*, Barlow and Clarke (2002), pp. 56-57, referring to the UN.

[15] Postel (1996), p. 13.

[16] Postel (1996), p. 15. See Gleick, Burns *et al.* (2002), p. 289, Table 10, on Irrigated Area, by Region, 1961 to 1999, and p. 290, Table 11, on Irrigated Area, Developed and Developing Countries, 1960 to 1999.

[17] WWAP (2003), p. 13.

[18] WWAP (2003), p. 192.

in India over-exploitation of aquifers could result in the loss of a quarter of its grain harvest in the near future.[19] Nevertheless, according to the WWDR, by 2050 all people could have access to food and undernourishment is attributable to 'global and national social, economic and political contexts that permit, and sometimes cause, unacceptable levels of poverty to perpetuate.'[20]

2.2.3   Cultural uses

In the majority of cultures, water plays an important part in religion as well as other human thought, activities or customs that are largely dictated by and part of the culture of a region. In religions such as Hinduism, Islam and Christianity, water is thought to have such functions as purifier, symbol of life and fertility and a means to fight evil.[21] The sacred character of water can be incompatible with human ownership.

Water is also used to a large extent in tourism and recreation. While in the 1970s one in thirteen people travelled from an industrial state to a developing country, by the end of the nineties one in five people did so. As a result, tourism has increasingly become an important source of income for many developing countries. Nevertheless, tourist consumption of water is high and adequate waste water treatment is lacking in most of these countries.[22] Much can be done in the tourist industry to increase the efficiency of fresh water use, such as through the re-use of towels by guests and by other hygiene regulations in hotel businesses.[23] As an industry, tourism falls within economic uses as well.

Recreation often requires a specific quality of water found in a healthy environment. For example, a lake that meets the required ecological standard will on average also be suitable for recreation such as swimming. On the other hand, pastimes such as golf, especially in arid areas, can be disastrous for the ecology of a region, as can the cultivation of lawns and growing oranges in a desert area. As stated by the WDDR: 'Golf tourism has an enormous impact on water withdrawals – an eighteen-hole golf course can consume more than 2.3 million litres a day.'[24]

---

[19] Barlow and Clarke (2002), p. 24, referring to the International Water Management Institute.
[20] WWAP (2003), p. 220.
[21] Disanayaka (2000).
[22] WWAP (2003), p. 16.
[23] Schachtschneider (2002).
[24] WWAP (2003), p. 16.

2.3   Economic uses

The uses categorised under the economic pillar of sustainable development
are those that mainly serve economic development. They comprise industrial uses, transport and energy.

2.3.1   Industrial uses

Industry, together with economic growth, has rapidly grown during the
nineteenth and twentieth centuries and is still growing fast. Water is used
in almost all industrial processes, from the processing of raw materials by
mining and cooling, the recycling of bottles, to the use of water in the
product itself, such as in sodas.

Industry is regarded as the second largest user of water, using about 22
percent of the total share, and is an important contributor to pollution as
well, while municipal use of water is estimated to account for about 6 percent of the total share.[25] These human and industrial waste products present a danger to human and aquatic health. For example, wastes containing
cadmium, lead and mercury can affect reproduction.[26]

The technical industry is often expected to solve water problems. On
the one hand, for example, the fast development of information technology has aided in the application of technical solutions such as offered by
remote sensing. On the other hand, technology often calls for more technology and, moreover, is not always the clean business it is supposed to be:
the computer industry in the US alone annually uses nearly 1,500 billion
litres of water and produces over 300 billion litres of wastewater in the
production of computer goods for which they need huge quantities of de-
ionised fresh water.[27] Technical solutions to the water crisis are therefore
to be thoroughly reviewed before implementation and cannot be depended
upon solely.

2.3.2   Transport

Transport on water mainly refers to navigation. Navigation encompasses
huge economic interests. About 90 per cent of trade in goods is estimated
to take place through shipping. For landlocked countries, navigation is also
important in order to gain access to the sea. On the other hand, shipments
can disturb ecological areas in various ways. For example, many alien spe-

---

[25] See Postel (1996), p. 14, Table 2 on Estimated Global Water Demand and Consumption, by Sector, c. 1990, based upon S.L. Postel, G.C. Daily, and P.R. Ehrlich, 'Human
Appropriation of Renewable Freshwater', *Science*, February 9, 1996.
[26] WWAP (2003), p. 15, furthermore stating: 'Industrial wastewater, like municipal
sewage, often contains suspended solids that silt up waterways, suffocate bottom dwell-
ing organisms and impede fish spawning.'
[27] Barlow and Clarke (2002), p. 8.

cies have been introduced in totally different biological ecosystems through navigation, destabilising the biotope. The fact that modern international law relating to freshwater resources tends to concentrate on the non-navigational uses of water does not imply the unimportance of navigation. On the contrary, its early regulation by international law has left relatively few areas of contention.[28]

Timber floating and rafts, practiced in northern Europe and the United States, is another use of water for transport, which can come into conflict with navigation.[29] The use of fresh water for transport can also consist of, for example, movement of coal through water pipes. Transport is, moreover, required in all cases in which water is moved from one location to another.

The increase in transport for trade and in water is expected to further damage waterways and ecosystems. In addition, the percentage of population living at distance from freshwater resources continues to increase, most likely resulting in growing transport of water. An example of an extremely unsustainable large city away from water resources is Las Vegas, which is located in a desert area but where water is abundantly used for luxury products and services.[30]

2.3.3 Hydropower

Energy production by hydropower is another important use of fresh water.[31] Other uses of water for energy include the use of water for cooling in the production of electricity, which is said to be relatively environmentally friendly except for the impact of temperature and water flow changes. Although water is needed for almost all energy production, hydropower is the main form of water use for energy production. One of the earliest recorded examples of a large dam is the Marib or Mareb dam in Yemen, believed to be constructed between 1000 and 500 BC and which collapsed around 600 AD.[32] The dam stored water for irrigation at the time of Queen Sheba and King Solomon.

Hydropower in 2001 accounted for about 19 percent of the world's total electricity production. Industrialised states have used about 70 percent of their hydroelectricity potential in contrast to about 15 percent by developing countries.[33] Hydroelectricity has resulted in great benefits. Hydropower is generated through the construction of dams.[34] Dams have

---

[28] Tanzi and Arcari (2001), pp. 49-51.

[29] See Caponera (1992), p. 214.

[30] Cf. Barlow and Clarke (2002), p. 238.

[31] See, e.g., Caponera (1992), pp. 214-215, on international cooperation in the field of hydroelectric power.

[32] See www.yementimes.com on the Mareb Dam.

[33] WWAP (2003), p. 13.

[34] See Gleick, Burns et al. (2002), pp. 291-295, Table 12, on Number of Dams, by Continent and Country, pp. 296-299, Table 13, on Number of Dams, by Country, p.

been constructed for different purposes such as prevention of floods and production of energy.[35] Dams serve to store and divert water for other uses, such as irrigation, as well.[36] Prosperity can often be linked to the construction of irrigation networks, such as formed by the ancient water reservoirs in the Anuradhapura District in Sri Lanka.[37] The number of dams in the world has increased enormously over the last fifty years, rising from just over 5,000 large dams in 1950 up to about 38,000 in 1996.[38]

More recently, the construction of, especially large, dams for hydropower or other uses has become increasingly controversial. Negative social and ecological impacts of such projects are becoming more and more known. It is estimated that between 60 and 80 million people have been displaced over the last 60 years because of the construction of dams. Such contentious issues related to large dams were addressed by the independent World Commission on Dams (WCD).[39] The WCD estimated that in India alone about 16 to 38 million people have been displaced because of large dams, while projects such as the construction of the Three Gorges Dam in China substantially increase the numbers of displaced people.[40] The relocation of people can coincide with violence, ranging from destruction of houses such as in China, to the killing of indigenous people such as in Guatemala.[41]

Dams can, moreover, cause drastic alteration of environmental circumstances affecting fishery and habitat.[42] The impact of dams on the flow of rivers is said to have resulted in altering three-quarters of the flow of major rivers in the northern hemisphere.[43] Although the pros and cons of dams are often debated, it is safe to say that especially large dams can threaten coastal ecosystems, wetlands and biodiversity.

An additional concern especially in relation to large dams is the fact that the costs of their maintenance and removal have often not been taken into account. The consequences of the construction of large dams and their

---

300, Table 14, on Regional Statistics on Large Dams, and pp. 301-302, Table 15, on Commissioning of Large Dams in the 20th Century, by Decade.

[35] Bertels *et al.* (1999), p. 130.

[36] Salman (2001b).

[37] Van der Molen (2001).

[38] Postel (1996), p. 29.

[39] See WCD (2000). The WCD report, *e.g.*, recommends the development of a rights-based and risk assessment approach (rather than cost and benefit) and includes strategic priorities, criteria and guidelines for decision-making toward an equitable and sustainable development of water and energy resources. See on the Commission and for the report www.dams.org.

[40] WCD (2000), p. 104.

[41] See Barlow and Clarke (2002), pp. 61-63.

[42] See Bertels *et al.* (1999), pp. 130-131, on negative impacts of dams, such as the Aswan dam, for example on fish populations.

[43] Barlow and Clarke (2002), p. 9.

limited life expectancy appear not to be fully envisaged.[44] While the construction of dams continues especially in developing countries, the decommissioning of dams appears to be the new trend in the US since 1998.[45]

2.4   Ecological uses

The uses categorised under the ecological pillar of sustainable development are those that relate to the natural functions of water. The need for water by the aquatic ecosystems, the hydrological cycle and water as part of the world ecosystem are now discussed.

2.4.1   Aquatic ecosystems

Water is a necessity for all life and therefore for all fauna and flora and especially for aquatic ecosystems.[46] Article 3.1 of the Berlin Rules defines "aquatic environment" as 'all surface waters and groundwater, the lands and subsurface geological formations connected to those waters, and the atmosphere related to those waters and lands.'[47] Aquatic ecosystems often require a higher water-table compared to agricultural and urban uses. According to Barlow and Clarke, freshwater systems are estimated to offer a home to 12 percent of the animal species: 'Yet over the last several decades, at least 35 percent of all fresh water fish species have become extinct, threatened, or endangered, and entire fresh water fauna systems have disappeared.'[48] Pollution of water is one of the main threats to aquatic ecosystems. As an extreme example, the reallocation of the waters of the Aral Sea for agriculture and industry has not only caused this vast inland sea to shrink by about three-fourths and caused the degradation of its environment, but has also led to a high level of toxic pollutants in the remaining water, while the number of species of fish has been reduced from 24 to 4, illustrating the rapid loss of aquatic biodiversity that can result from human intervention.[49] The impact of human activities contributes, for example, to losses of sedimentary material in the ocean and receding coastlines, and to losses of wetlands and other habitats, both in their turn influencing coastal ecosystems and biodiversity.[50]

The present withdrawal of water is in many cases endangering the environment, possibly compromising the needs of present generations, let

---

[44] An example of such a possible effect is that the weight of the water held by dams may cause earth tremors. See Barlow and Clarke (2002), p. 49.

[45] See, *e.g.*, Barlow and Clarke (2002), p. 202.

[46] See on freshwater ecosystems, *e.g.*, J.N. Abramovitz (1996).

[47] See on the 2004 ILA Berlin Rules note 60, Section 3.3.1.

[48] Barlow and Clarke (2002), p. 27.

[49] Postel (1996), p. 7, continuing to state that fish catch used to total 44,000 tons a year, supporting 60,000 jobs. See on the Aral Sea also De Villiers (2001), pp. 105-116.

[50] Bertels *et al.* (1999), pp. 130-131.

alone those of future generations.[51] The rate at which changes are taking place often does not provide sufficient opportunity for adaptation by the ecosystems. Environmental degradation often has adverse effects on, for example, flood control, especially in the case of the disappearance of wetlands, which provide natural storage for water, or deforestation, destroying the natural soil cover and the root systems that keep the soil together. The complicated effects of degradation of aquatic ecosystems can be irreversible. Prevention is therefore of all the more importance.

2.4.2   The hydrological cycle

The hydrological cycle is the constant movement of water through a cyclic process.[52] Water evaporates from the oceans (about 430,000 km3/yr), the main part of which, returns to the sea (about 390,000 km3/yr). About 40,000 km3/yr of evaporation from the sea reaches land. Precipitation on land territory is about 110,000 km3/yr, of which about 70,000 km3/yr evaporates or is transpired by plants and about 40,000 km3/yr runs off to the sea. An even smaller part can be found in space. Of the precipitation on land, part flows into surface water, part goes underground, and part is taken up by fauna and flora. Groundwater referred to as "confined" is captured between layers of rock or hard sediment, but is rarely really contained. Although the pace of replenishment can be extremely slow, even confined water resources are part of the cycle in the really long-term.

Since virtually all freshwater resources are part of this system, all uses have an impact on the system as a whole and they are, in turn, all influenced by the hydrological cycle. In other words, water flowing to the sea is not wasted. The hydrological cycle presents a complicated process of interconnectedness, influenced by actions such as large diversion of water. Despite modern technology including satellites and extensive research, information on aquifers remains insufficient. Their location, their connection to surface water and their replenishment rate are often not (fully) known. Therefore, the exact impact of groundwater pollution and water withdrawal is hard to establish. For example, depending on the underground flow, withdrawal in one country can lead to lower water-tables in a neighbouring country. Moreover, it can take centuries before the consequences of pollution of groundwater come to surface.

2.4.3   The world ecosystem

Through the hydrological cycle, water is an essential factor in climatic systems in relation to the world's ecosystems, from forests to deserts. Water,

---

[51] Gleick, Burns *et al.* (2002), pp. 243-251, Table 2, on Fresh Water Withdrawals, by Country and Sector.
[52] See also Chapter 4 of WWAP (2003).

climate, forests, deserts, fauna and flora are all part and parcel of a single, global ecosystem.

The relation between the hydrological cycle and climate change is complicated and manifold and small changes can already strongly influence water balances.[53] The climate is highly influenced by the main ocean water flows and, *vica versa*, the ocean winds influence the water flows.[54] Climate change influences regional water availability and is likely to cause extremes in both rainfall and droughts. It can moreover change the temperature of freshwater lakes and ocean surface water, affecting in turn the related flora and fauna. Climate change may cause the sea-level to rise and alteration of coastal areas and tidal rivers. A rise with concomitant sea-level could cause or contribute to, for example, the endangering of small island states. It could, moreover, cause further saltwater intrusion of coastal areas. The drastic influence of changes in the temperature of water on the polar biotope is one of the main underlying concerns surrounding the melting of polar ice. Climate change could furthermore pose a threat to wetlands, whose loss would mean a loss of habitats, and consequently of biodiversity, especially in conjunction with the over-exploitation of species.

Biological diversity (biodiversity) is part of ecology and refers to, *inter alia*, the number and variety of living organisms. Biodiversity is threatened by the rate at which loss of species takes place, outrunning the evolution of species.[55] Biodiversity is also threatened by deforestation. Deforestation reduces the ability of land to retain water and therefore leads to an increased risk of erosion and excessive floods. For example, rainforests absorb large parts of seasonal rain, protecting land such as surrounding the Amazon river. Forests furthermore regulate local and regional climates by their recycling of water. Deforestation has an impact on water quantity as well as on water quality. Acid rain, which can contaminate lake waters, is also a threat to forests because it alters the acidity of the soil. The availability of water is furthermore a major factor in the process of desertification, which is said to cover 3.6 billion hectares in over a hundred states.[56]

---

[53] See on climate change, *e.g.*, Gupta (2001) and Bertels *et al*. (1999). See on the relationship between international water law and climate change, Tarlock (2000). See on human weather modification, *e.g.*, Davis (1991).

[54] See, *e.g.*, Dellapenna (2001), p. 244: 'Evaporation from the sea is the great engine that drives weather across the planet.'

[55] Extinction rates are estimated to be one hundred to one thousand times higher than in the pre-human era. See Barlow and Clarke (2002), p. 27, citing *Science*. See also Van der Zwaan and Petersen (2003) on the relation between population growth, human consumption and loss of species.

[56] Barlow and Clarke (2002), p. 45, citing the FAO.

2.5 Conclusions

The proportion of fresh water allocated to a particular use differs from one region to another: the largest part in developed countries is devoted to industrial uses, whereas in developing countries agriculture is the main use.[57] In all regions, a certain amount and quality of water is to be safeguarded for domestic uses in order to meet basic human needs. The tremendous amounts of water used for, and polluted by, food production often conflict with other uses of water. Over-exploitation of aquifers can, for example, reduce the access to drinking water. The degradation of the environment by pollution also has its impact on fisheries, considering that a huge part of the fish used for human consumption spend part of their life cycles in wetlands and estuaries.[58] Another conflict within food production concerns the competing interests over water of settled agricultural communities and travelling nomads practicing pastoralism, such as in the horn of Africa.

Recreation and tourism are other social uses of water that often conflict with environmental interests. Tourism can easily conflict with uses for the domestic population: the money it generates can inflate the price of fish and finance swimming pools in water-stressed areas while the local population's needs – for drinking water, sanitation and nutrition – are not fulfilled. However, under certain conditions, recreation and environment can be a relatively sustainable and profitable combination as well.

The worldwide growth of industry is another important cause of the increasing pollution of fresh water. Within modern industry, processes for the efficient use of water, such as a closed system of water use, are increasingly being developed. Combining economic uses with social and ecological uses is difficult: commercial undertakings on average aim at the highest possible profit and continuing growth, whereas social concerns refer to equitable use of water and ecological interests call for decreased consumption.

In setting priorities for freshwater use, ecosystems are still mainly looked upon through the lens of human benefits and costs. However, the urgency of the need for proper ecological management is becoming increasingly obvious. Freshwater resources that satisfy environmental criteria also usually fulfil the quality requirements of other uses, further encouraging serious consideration of ecological interests.

The need to take all of the aforementioned interests into account can be put to practice by means of the concept of Integrated Water Resources Management (IWRM), aimed at achieving sustainable development.[59] According to the WWAP: 'IWRM is based on the understanding that water

---

[57] Bertels *et al.* (1999), pp. 129-130, referring to a report of the World Resources Institute (WRI) of 1992.

[58] Bertels *et al.* (1999), p. 131, referring to WRI, 1994.

[59] See WWAP (2003), pp. 376-377, and Savenije and Van der Zaag (2002).

resources are an integral component of the ecosystem, a natural resource and a social and economic good.'[60] At least at the international political level, the IWRM approach seems to have first been formulated at the 1992 International Conference on Water and the Environment in Dublin and the resulting Dublin Statement.[61] IWRM 'can be regarded as the vehicle that makes the general concept of sustainable development operational for the management of freshwater resources.'[62]

Whilst acknowledging the unity of the global hydrological cycle, geographic characteristics of freshwater resources differ significantly, mainly from one region to another. Likewise, social and economic uses and interests vary widely, even between different communities and at the national level. Therefore, IWRM should preferably be organised per drainage basin, or catchment area.[63]

The potential conflicts between uses and users of water are numerous and severe. The allocation of fresh water between its various uses has become a major issue at the community, national as well as the international level. At the level of states, conflicts have often risen, for example, between the interests of upstream and downstream states.

The problems related to fresh water and the interaction between the world's waters affect the required response of international water law and call for an integrated approach to achieve sustainable development. The next Chapter introduces the existing and emerging international law relevant to fresh water and its ability to contribute to the sustainable development of freshwater resources.

---

[60] WWAP (2003), p. 377.
[61] See Chapter 1, note 23, of this study.
[62] WWAP (2003), p. 37.
[63] The catchment basin approach is discussed in Section 6.4.1 of this study.

3. Sustainable development in international law

3.1 The status of sustainable development

The status of sustainable development under international law is determined by the sources of international law. The commitment to sustainable development expressed by the international community, such as through the Rio Declaration, is to be found incorporated in various multilateral treaties. These include treaties on climate change, biological diversity, desertification, watercourses, the treaty establishing the WTO, and regional treaties such as the ECE treaties, the EC Treaty and the EU Constitution.[1]

The ICJ explicitly referred to the concept of sustainable development in its 1997 judgement in the *Gabčíkovo-Nagymaros* case (Hungary/Slovakia) on the Danube river.[2] In its judgement, the ICJ stated that 'This need to reconcile development with protection of the environment is aptly expressed in the concept of sustainable development.'[3] In the *Gabčíkovo-Nagymaros* case it was, moreover, stated that the implementation of a treaty is not a static matter but has to take into account the emergence of new norms, such as of environmental law. In his Separate Opinion to the judgment Vice-President Weeramantry elaborates more thoroughly on the concept of sustainable development and argues strongly that sustainable

---

[1] See, *e.g.*, ILA Committee on Legal Aspects of Sustainable Development (2002), para. III, on the incorporation of sustainable development in various treaties and references to it in international judicial decisions, and Gupta (2003), on sustainable development in the UNFCCC. For the Constitutional Treaty for Europe as agreed by EU leaders at the European Council, Brussels, 17 and 18 June 2004 and to be signed on 29 October 2004 in Rome, see http://europa.eu.int/futurum/eu_constitution_en.htm. See also europa.eu.int/comm/sustainable/index_en.htm.

[2] International Court of Justice (1997), 25 September 1997, *Case Concerning the Gabčíkovo-Nagymaros Project (Hungary/Slovakia)*. The Judgment and Opinions can be found on www.icj-cij.org. The case concerns certain issues arising out of differences regarding the implementation and the termination of the construction and operation of the Gabčíkovo-Nagymaros barrage system, a system of locks in the Danube, as agreed between Hungary and Czechoslovakia by the Budapest Treaty of 16 September 1977. The Court found both parties in breach of their legal obligations. Hungary by suspending and abandoning the project (alleging that it entailed grave risks to the Hungarian environment and the water supply of Budapest) and Slovakia (state-successor of Czechoslovakia) by putting into operation an alternative project ('Variant C') that affected Hungary's access to the water of the Danube. The 1977 treaty was considered still in force and the parties were called by the ICJ to negotiate in good faith. Since negotiations did not result in a solution, Slovakia filed a request for an additional judgement on 3 September 1998, asking the Court to determine the modalities for executing the judgement. The case is still pending before the Court. See ILA Committee on Legal Aspects of Sustainable Development (2000), pp. 13-14, and Stec (1999).

[3] *Gabčíkovo-Nagymaros* Judgment, para. 140.

development is not merely a concept, but a principle that is now accepted worldwide. Although much more far-reaching than the actual judgement itself, this Opinion is of interest because of its progressive perspective. Judge Weeramantry states that the principle of sustainable development is 'a part of modern international law by reason not only of its inescapable logical necessity, but also by reason of its wide and general acceptance by the global community.'[4] According to the ILA Committee on Legal Aspects of Sustainable Development: 'sustainable development has become an established objective of the international community and a concept with some degree of normative status in international law.'[5]

Some experts argue that sustainable development is not just a legal concept but an evolving body of international law – the international law of sustainable development.[6] The second perspective is taken as a starting point in 3.2. International water law and the extent to which sustainable development and related principles are embedded in present international water law are reviewed in 3.3. In 3.4 those trends that need to be further encouraged for international water law to contribute to the sustainable development of freshwater resources are identified. As a resultant of the methodology of this study, those principles of international law identified as potentially contributing to the achievement of sustainable development in freshwater management are further identified and explored in the relevant parts of following chapters.

3.2   International law on sustainable development

The emerging international law on sustainable development can provide guidance for the further evolution of international water law toward sustainable development.[7] The ILA is an academic authority in the field of international law and its New Delhi Declaration on Sustainable Development reflects the emerging shape of international law on sustainable development.[8] The Declaration is a codification attempt of the ILA Committee

---

[4] Separate Opinion of Vice-President Weeramantry to the *Gabčíkovo-Nagymaros* Judgment, under para. A(c) at p. 5.

[5] ILA Committee on Legal Aspects of Sustainable Development (2002), p. 5.

[6] *Cf.* Gupta (2004), pp. 32-34.

[7] On international law and sustainable development, see ILA Committee on Legal Aspects of Sustainable Development (2002) and the First Report of the ILA Committee on International Law on Sustainable Development (2004), Schrijver and Weiss (2004), Boyle and Freestone (1999) and Sands (1995). The ILA Committee on International Law on Sustainable Development, formed in May 2003, builds upon the work of the Committee on Legal Aspects of Sustainable Development that completed its work in 2002, and is planning to study, *e.g.*, the status and implementation of the principle of integration, developing States in a changing global order and selected aspects of the international law of development.

[8] See www.un.org/ga/57/document.htm for the ILA New Delhi Declaration.

on Legal Aspects of Sustainable Development and – considering the involvement of highly qualified publicists – is arguably a secondary source of international law.[9] The ILA New Delhi Declaration identifies seven main principles of international law, relevant to the activities of all actors involved, whose application and, where relevant, consolidation and further development would be instrumental in effectively ensuring sustainable development (Preamble). The seven principles are: the duty of states to ensure sustainable use of natural resources;[10] the principle of equity and the eradication of poverty; the principle of common but differentiated responsibilities; the principle of the precautionary approach to human health, natural resources and ecosystems; the principle of public participation and access to information and justice; the principle of good governance; and the principle of integration and interrelationship, in particular in relation to human rights and social, economic and environmental objectives. These principles will be referred to throughout the various relevant parts of this study.

The international law on sustainable development emerged from international environmental law and the international law on development. Whether or not the body of human rights law is a third main source of the law on sustainable development has been subject of discussion. As reflected in the ILA New Delhi Declaration, the essential relevance of human rights to sustainable development requires it to be integrated into the concept.[11] In broad terms, human rights can be regarded as mainly relating to the social pillar, developmental law as relating to the economic pillar and environmental law to the ecological pillar of sustainable development.

3.2.1 Human rights

As stated in the ILA New Delhi Declaration: 'the realization of the international bill of human rights, comprising economic, social and cultural rights, civil and political rights and peoples' rights, is central to the pursuance of sustainable development'. The concept of human rights has progressed significantly since World War II.[12] The 1948 Universal Declaration of Hu-

---

[9] ILA Committee on Legal Aspects of Sustainable Development (2002).

[10] The duty to ensure sustainable use of natural resources refers to but is not necessarily limited to states.

[11] On the relationship between human rights, democracy and development see Arts (2000), pp. 24-26.

[12] Under the Charter of the United Nations the UNGA can initiate studies and make recommendations regarding human rights (Article 13). The UN shall promote universal respect for, and observance of, human rights and fundamental freedoms for all (Article 55). For the achievement of this purpose, all members pledge themselves to take action (Article 56). ECOSOC may also make recommendations on human rights, draft conventions for the Assembly and call international conferences on human rights matters (Article 62).

man Rights (UDHR) contains various rights and freedoms.[13]

In 1966 the UDHR was elaborated in the following two treaties: the International Covenant on Economic, Social and Cultural Rights (ICESCR) and the International Covenant on Civil and Political Rights (ICCPR).[14] This separation into two treaties reflected the division in the Cold War period between "the West" and "the East". Another way of categorising human rights is by referring to the different generations of rights. Civil and political rights - or first generation rights - are mainly found in the ICCPR. These include: the right to life, prohibition of torture, prohibition of slavery, the right to liberty and security of person, the right of liberty of movement and residence within territory, equality before courts, the right of recognition as a person before the law, the right of privacy, freedom of thought, conscience and religion, freedom of expression and freedom of association. The rights under the ICCPR are formulated in such a way as to create obligations that states must respect and ensure (Article 2). These rights often can be immediately implemented since they in principle require non-intervention rather than intervention by a government.[15] In line with the liberal notion that a government should primarily provide for the freedom of its citizens by non-intervention, these rights are mainly favoured by the US, European and other western countries. The Human Rights Committee ensures the implementation of the ICCPR by state parties. The Committee monitors a reporting system and can make general comments. State parties may recognise the competence of the Committee to hear interstate complaints and by an optional protocol this Committee can receive and consider individual communications on violations by a state party.

Economic, social and cultural rights – also referred to as second generation rights – are primarily found in the ICESCR.[16] These rights include: the right to work, the right to social security, the right to an adequate standard of living, the right to health, the right to education and the right to take part in cultural life. Under this covenant, states are obliged to take steps as far as their available resources allow them to progressively achieve the realisation of these rights (Article 2). This formulation gives rise to

---

[13] Universal Declaration of Human Rights, adopted and proclaimed by UNGA resolution 217 A (III), 10 December 1948, *UN Doc.* A/810 (III).

[14] International Covenant on Economic, Social and Cultural Rights, adopted by UNGA Resolution 2200 A (XXI), New York, 16 December 1966, entry into force: 3 January 1976, 993 *UNTS*, 3, and 6 *ILM* (1967), 360. Status as of 4 October 2004: 150 parties and 65 signatories. International Covenant on Civil and Political Rights and Optional Protocol, adopted by UNGA Resolution 2200 A (XXI), New York, 16 December 1966, entry into force: 23 March 1976, 999 *UNTS*, 171, and 6 *ILM* (1967), 368. Status as of 4 October 2004: 153 parties and 67 signatories.

[15] However, in practice their implementation, *e.g.* of the right to vote, often requires active intervention of and facilitation by a government.

[16] See Steiner and Alston (2000) on human rights developments during the Cold War and on economic, social and cultural rights.

obligations that can be gradually implemented, depending on resources. These rights were preferred in the former Soviet Union and are still favoured in countries that prioritise economic development, such as China and many other Asian countries as well as developing countries in other regions. The Committee on Economic, Social and Cultural Rights monitors the implementation of these rights. The position of this Committee is not as strong as that of the Human Rights Committee. Apart from its reporting procedures, the Committee cannot hear individual petitions, nor is there an interstate complaints procedure, although proposals for such procedures have been made. Yet, it is now widely accepted that economic, social and cultural rights form a firm part of the indivisible universal human rights package.

Collective rights – or third generation rights – concern rights that cannot be fully implemented per person since they concern a group of people, such as rights for the protection of a language or culture or group of people. Apart from the primordial right of self-determination of peoples expressed in Article 1 of the 1966 treaties, collective rights are scarcely found in those treaties. In contrast, many collective rights are included in the African Charter on Human and Peoples' Rights.[17] This reflects the appreciation of these rights by many developing countries and the scepticism of industrialised countries in relation to these rights.[18]

Although the importance of human rights is widely accepted, confusion remains on their precise nature and role in international law.[19] In the *Barcelona Traction* case, the ICJ stated that obligations derived from, for example, the principles and rules concerning the basic rights of the human person are by their very nature the concern of all states.[20] The ICJ continued: 'In view of the importance of the rights involved, all States can be

---

[17] African Charter on Human and Peoples' Rights, adopted by the 18th Ordinary Session of the Assembly of Heads of State and Government of the Organisation of African Unity, now transformed into the African Union (AU), Nairobi, 27 June 1981, entry into force: 21 October 1986, 21 *ILM* (1982), 59. All 53 AU Member States are state parties. Collective rights included in this Charter are the right of all peoples to equality (Article 19), the right to existence (Article 20), the right to freely dispose of their wealth and natural resources (Article 21), the right to their economic, social and cultural development (Article 22), the right to national and international peace and security (Article 23) and the right to a general satisfactory environment (Article 24).

[18] See, *e.g.*, the Netherlands Advisory Council on International Affairs, *A human rights based approach to development cooperation*, report No. 30, April 2003, www.aiv-advies.nl, in which at p. 16 reference is made to the debate on collective rights. In this report, one of the presumptions is the indivisibility of political and civil, social, economic and cultural rights as well as rights of peoples.

[19] Shaw (1997), pp. 196-198.

[20] ICJ Judgement in the *Case concerning the Barcelona Traction, Light and Power Company Limited* (Belgium/Spain), ICJ Rep. 1970, p. 32. See also ICJ, *Reservations to the Convention on the Prevention and Punishment of the Crime of Genocide*, 28 May 1951, ICJ Rep. 1951, p. 23: 'the contracting States do not have any interest of their own; they merely have, one and all, a common interest'.

held to have a legal interest in their protection; they are obligations *erga omnes*.'[21] These statements were affirmed by the ICJ in its Advisory Opinion on the construction of a wall by Israel on Palestinian territory.[22] A human right can, moreover, be *jus cogens*, and therefore a peremptory norm of general international law from which no derogation is permitted.[23]

Post-Cold War treaties better reflect the interrelationship of the generations of human rights, for example the 1989 UN Convention on the Rights of the Child (CRC).[24] The universality of human rights has been the subject of discussion, but tends to be increasingly accepted.[25] As expressed in the 1993 Vienna Declaration:[26]

> All human rights are universal, indivisible and interdependent and interrelated. The international community must treat human rights globally in a fair and equal manner, on the same footing, and with the same emphasis. While the significance of national and regional particularities and various historical, cultural and religious backgrounds must be borne in mind, it is the duty of States, regardless of their political, economic and cultural systems, to promote and protect all human rights and fundamental freedoms.

The crucial nature of access to water – whether for survival or for the development of a people – cuts across the categories of human rights and

---

[21] ICJ *Barcelona Traction* case, p. 32. Human rights with an *erga omnes* character include those formulated in the 1948 Genocide Convention, see ICJ Judgement in the *Case on Application of the Convention on the Prevention and Punishment of the Crime of Genocide*, Preliminary Objections (Bosnia and Herzegovina/Yugoslavia), ICJ Rep. 1996, para. 31.

[22] ICJ Advisory Opinion on *Legal Consequences of the Construction of a Wall in the Occupied Palestinian Territory*, 9 July 2004, General List No. 131, para. 155, www.icj-cij.org.

[23] See on *jus cogens*, Article 31 of the 1969 Vienna Convention on the Law of Treaties. An example is the right to be free from torture.

[24] See Arts (2000), p. 22. The Convention on the Rights of the Child was adopted by the General Assembly, New York, 20 November 1989, Resolution 44/25, 1577 *UNTS*, 3, entry into force: 2 September 1990. As of 4 October 2004: 192 parties, which is all states except for the US and Somalia who are the 2 remaining signatories. See also www.unicef.org/crc.

[25] However, the debate between universalism and relativism remains alive on interpretation and priorities of the various human rights. See on this debate, *e.g.*, Arts (2000), pp. 31-36, who states at p. 34: 'Whereas human rights law on the one hand affirms the universal character of human rights, on the other hand it clearly allows for exceptions.' Defining the exceptions seems to be the focus of the discussion.

[26] Chapter I, para. 5, of the 1993 Vienna Declaration of the World Conference on Human Rights, *UN Doc.* A/CONF.157/23, adopted on 25 June 1993. See also the Proclamation of Teheran, proclaimed by the International Conference on Human Rights at Teheran on 13 May 1968, para. 13: 'Since human rights and fundamental freedoms are indivisible, the full realization of civil and political rights without the enjoyment of economic, social and cultural rights is impossible.'

underlines the actual indivisible nature of human rights. The position of access to water within the body of human rights law is further reviewed in Section 4.2 of this study, considering its important role in water as a social good.

3.2.2   International development law

International development law can be viewed as 'an instrument for the economic and legal transformation of international relations and as a means for giving all States an opportunity to take part in international life on a footing of true equality.'[27] International development law does, however, not only concern the relations between states. In the 1986 UNGA Declaration on the Right to Development (DRD), development has been defined as:[28]

> a comprehensive process, which aims at the constant improvement of the well-being of the entire population and of all individuals on the basis of their active, free and meaningful participation in development and in the fair distribution of benefits resulting therefrom.

International development law relates to the eradication of poverty, permanent sovereignty over natural resources, entitlement to development assistance and transfer of technology, preferential treatment of developing countries in trade and common but differentiated responsibilities.[29] As a resultant of the methodology, various principles of international development law that are of great significance to sustainable development of freshwater resources are discussed in Parts II and III of this book. The right to development, the right of self-determination and the principle of common but differentiated responsibilities are elaborated in Chapter 7 since they are viewed to represent the combination of the social and economic pillars.

In contrast to the seventies and eighties of the 20th century in which there was a search mainly by developing countries for a New International Economic Order of which international development law was part and parcel, international development law seems to have received only little

---

[27] Bulajić (1993), p. 43. See, *e.g.*, Sen (1999) on development. On relevant international economic law, including the WTO, see Section 5.4 of this study.

[28] UNGA (1986), Declaration on the Right to Development, *UN Doc*. A/RES/41/128, preambular para. 2. The DRD was adopted by 146 votes to 1 (USA) and 8 abstentions (Denmark, FRG, Finland, Iceland, Israel, Japan, Sweden, UK).

[29] See on international development law, *e.g.*, Schrijver (2001), Bulajić (1993), De Waart, Peters and Denters (eds) (1988) and 'Progressive Development and Norms of International Law relating to a New International Economic Order: Report of the Secretary-General', *UN Doc*. A/39/504/Add. 1, 23 October 1984. See on transfer of technology also, *e.g.*, Hedger, Natarajan, Turkson and Wallace (2000) and Li (1994).

attention in the last few decades.[30] Although its present status within international law can be debated, the need for such development law is in no
way outdated. As concluded by Schrijver: 'Particular concerns are the continued conflicts of interests between developing and industrialized States
and the question whether, and to what extent, developing States have a
discretion to determine their own development and environmental policies
in an era of globalization.'[31]

The link between water resources and development is well illustrated
by its positioning in the WSSD Plan of Implementation. Access to water in
the Plan of Implementation is primarily dealt with under Chapter II, on
poverty eradication, and under Chapter IV, on the protection and management of the natural resource base of economic and social development.
For development, *e.g.* through the eradication of poverty, it is essential to
implement the goal of providing all people with sufficient quantities of
potable water. The position of developing countries and economies in transition needs special consideration, particularly in the case of the least developed countries and those countries with geographical disadvantages in
relation to fresh water such as countries threatened by desertification,
flooding or salinisation. In many circumstances they may need additional
assistance, including financial means, the transfer of technology and (other)
access to information.[32] Principle 6 of the Rio Declaration states:

> The special situation and needs of developing countries, particularly
> the least developed and those most environmentally vulnerable, shall
> be given special priority. International actions in the field of environ-
> ment and development should also address the interests and needs of
> all countries.

The relationship between water and development is well-formulated in
Agenda 21, Chapter 18, paragraph 18.6: 'The extent to which the development of water resources contributes to economic productivity and social
well-being is not usually appreciated, although all social and economic
activities rely heavily on the supply and quality of freshwater.' Moreover,
the cost of providing such access to water is far outweighed by the cost of
not doing so, mainly because of its effect on loss of productivity.[33]

---

[30] Development as such was the subject of, for example, the 2001 WTO Ministerial
Conference in Doha, see www.wto.org, and the International Conference on Financing
for Development, Monterrey, Mexico, 18-22 March 2002, see for the Report of the
Conference *UN Doc.* A/CONF.198/11.

[31] Schrijver (2001), pp. 25-26.

[32] According to Agenda 21, in order to be able to reach its developmental and environmental objectives, substantial financial resources are required for, among others, developing countries. These have not materialised so far.

[33] ACUNS, *Plan of Action for Johannesburg: The Development-Environment Nexus*,
distributed at PrepCom III, UN Headquarters, 28 March 2002.

3.2.3 International environmental law

From the late 60s and the 70s onward, environmental awareness and the protection of natural resources became more prominent in international law.[34] These concerns are expressed in conventions such as the 1968 African Convention on the Conservation of Nature and Natural Resources (signed in Algiers), the 1971 Convention on Wetlands of International Importance (Ramsar Convention) and the 1972 Convention for the Protection of the World Cultural and Natural Heritage.[35]

At present, international environmental law includes aspects of environmental pollution, fisheries, the law of the sea, the regulation of hazardous waste and nuclear energy, protection of climate, the atmosphere and biodiversity, the conservation of ecosystems, migratory species and marine living resources. The international law of the sea has to a large extent become integrated in the 1982 United Nations Convention on the Law of the Sea (UNCLOS).[36] Marine pollution is also subject of the 1992 OSPAR Convention.[37] International legal responses to the threats posed by climate change include the UN Framework Convention on Climate Change (UNFCCC), the Kyoto Protocol to the UNFCCC, the Vienna Convention for Protection of the Ozone Layer, and the Montreal Protocol on Substances that Deplete the Ozone Layer and its amendments.[38] Regulation of

---

[34] See on international environmental law, Birnie and Boyle (2002), Hunter, Salzman and Zaelke (2002) and Sands (1993).

[35] African Convention on the Conservation of Nature and Natural Resources, Algiers, 15 September 1968, entry into force: 16 June 1969, 1001 *UNTS*, 3. Convention on Wetlands of International Importance, Ramsar, 2 February 1971, entry into force: 21 December 1975, 996 *UNTS* 245, and 11 *ILM* (1972), 963, amended versions 1982 and 1987, entry into force: 1 May 1994. Convention for the Protection of the World Cultural and Natural Heritage, Paris, adopted by the General Conference of UNESCO at its seventeenth session on 16 November 1972, entry into force: 17 December 1975, see whc.unesco.org/wldrat.htm.

[36] UN Convention on the Law of the Sea (UNCLOS), Montego Bay, 10 December 1982, entry into force: 16 November 1994, 21 *ILM* (1982), 1261. Status as of 4 October 2004: 145 parties and 157 signatories. See Kwiatkowska, Dotinga, Molenaar, Oude Elferink and Soons (2002), Churchill and Lowe (1999), and Section 6.4.2 of this book.

[37] The Convention for the Protection of the Marine Environment of the North-East Atlantic (OSPAR Convention), Paris 1992, entry into force: 25 March 1998, replaces the Oslo and Paris Conventions. See www.ospar.org. The OSPAR Convention aims to prevent and eliminate all sources of marine pollution in the North-East Atlantic, including pollution from land-based sources, dumping and incineration.

[38] United Nations Framework Convention on Climate Change (UNFCCC), New York, 9 May 1992, entry into force: 21 March 1994, 1771 *UNTS* (1994), I-30822, 164-190, and 31 *ILM* (1992), 849. Status as of 4 October 2004: 189 parties and 165 signatories. The UNFCCC aims to achieve stabilisation of greenhouse gas concentrations in the atmosphere at a level that would prevent dangerous anthropogenic interference with the climate system (Article 2). See Gupta (1997) and Biermann (1995). The Kyoto Protocol to the UNFCCC, Kyoto, 11 December 1997, elaborates on obligations of the parties in achieving its quantified emission limitation and reduction commitments.

acid rain includes the 1979 Long-Range Transmission of Air Pollutants Convention and its Protocols. International legal responses to desertification include the 1977 UN Conference on Desertification at Nairobi and the 1994 UN Convention to Combat Desertification in Countries Experiencing Serious Drought and/or Desertification, Particularly in Africa (UNCCD).[39] The UN Convention on Biological Diversity was concluded in 1992 with the stated aim of conserving biological diversity and promoting both the sustainable use of its components and the fair and equitable sharing of its benefits.[40]

In addition, environmental considerations were expressed through conferences such as the 1972 UN Conference on the Human Environment (UNCHE) which lead to the establishment of the United Nations Environment Programme (UNEP).[41] International environmental law has, moreover, developed through UNGA resolutions, the work of international organizations such as the 1978 UNEP Principles on Shared Natural Resources and the work of NGOs such as the ILA Montreal Rules.[42] Serious warnings on the state of the environment, such as presented in 1987 by the

---

Status as of 5 October 2004: 126 parties, 84 signatories, and 44.2% of emissions. The Protocol is not yet in force since the parties need to account for at least 55% of emissions, which will be the case when Russia ratifies the treaty as soon expected. See *e.g.*, Faure, Gupta and Nentjes (2003). The Vienna Convention for the Protection of the Ozone Layer, Vienna, 22 March 1985, entry into force: 22 September 1988, 26 *ILM* (1987), 1529, obliges the parties to take appropriate measures to protect human health and the environment against adverse effects resulting from human activities which (are likely to) modify the ozone layer. Status as of 4 October 2004: 189 parties and 28 signatories. The Montreal Protocol on Substances that Deplete the Ozone Layer, 16 September 1987, includes special treatment for developing states and regard to the position of non-parties.

[39] UN Convention to Combat Desertification in Those Countries Experiencing Serious Drought and/or Desertification, Particularly in Africa (UNCCD), Paris, 17 June 1994, entry into force: 26 December 1996, 33 *ILM* (1994), 1328. Status as of 4 October 2004: 191 parties and 115 signatories. According to Article 2 of the UNCCD, the aim is to combat desertification through action at all levels and 'in the framework of an integrated approach consistent with Agenda 21, with a view to contributing to the achievement of sustainable development in affected areas.' See on the Convention www.unccd.int.

[40] Article 1 of the Convention on Biological Diversity, Rio de Janeiro, 5 June 1992, entry into force: 29 December 1993, 31 *ILM* (1992), 818. Status as of 4 October 2004: 188 parties and 168 signatories. See www.biodiv.org. See also the 2001 International Treaty on Plant Genetic Resources for Food and Agriculture, adopted by Res. 329/01 on the FAO Conference, 31st Sess., www.fao.org.

[41] UNEP was established as a subsidiary body by means of UNGA Res. 2997 (XXVII) (1972). On the UNCHE see Birnie and Boyle (2002), pp. 38-40. See also the Declaration of the UN Conference on the Human Environment, Stockholm 1972, *UN Doc.* A/CONF/48/14/REV.1. On UNEP see Birnie and Boyle (2002), pp. 53-54.

[42] The UNEP Principles include a principle on environmental impact assessment, see Section 6.2.3 of this study. 1982 ILA Montreal Rules on Transfrontier Pollution, Report of the 60th Conference (1982), 1.

Brundtland Commission, together with actual disasters including sinking ships that have caused massive pollution by oil spills, have furthermore led to the negotiation of treaties in various fields of the environment. In 1992, the conventions on climate change, biodiversity and desertification resulted from UNCED. Although a multilateral treaty on forests has not been reached, Forest principles were concluded at UNCED.[43]

International water law overlaps with international environmental law. Treaties providing for the environmental protection of water resources include the 1995 Agreement on the Cooperation for the Sustainable Development of the 1995 Mekong River Basin (Mekong Agreement) and the 1995 Protocol on Shared Watercourse Systems in the Southern African Development Community (SADC) Region.[44]

Principles of international environmental law relevant to water management include the well-established no-harm principle and other less established principles such as the right to a healthy environment, precautionary principle, polluter and user pays principle, and common heritage or concern of humankind. The right to a healthy environment, the precautionary principle and ecological considerations combined with the principle of justice are elaborated upon in Chapter 7 since they are viewed to represent the combination of the social and ecological pillars. The polluter and user pays principle, the no-harm principle and the common heritage or concern of humankind are also discussed in Chapter 7 since they bridge the economic and ecological pillars of sustainable development.

The formulation to 'take all appropriate measures', which is part of for example the obligation not to cause significant harm, indicates that the

---

[43] Non-legally binding Authoritative Statement of Principles for a Global Consensus on the Management, Conservation and Sustainable Development of All Types of Forests, UNCED Report, A/CONF.151/26 (Vol. III), 14 August 1992. Principle 4 states that the vital role of forests in, for example, protecting watersheds and freshwater resources, should be recognised.

[44] Agreement on the Cooperation for the Sustainable Development of the Mekong River Basin, Chiang Rai, 5 April 1995, between Cambodia, Laos, Thailand, Vietnam. See Browder and Ortolano (2000) on the Mekong regime, including an analysis of the Mekong Agreement. The Mekong Agreement is a regional framework of cooperation for sustainable development, utilization, conservation and management of the Mekong River Basin water and related resources for navigational and non-navigational purposes. The Mekong Agreement includes reasonable and equitable utilization and an integrated approach at basin level. The Mekong River Commission is the institutional framework. Riparian states are the addressees. The aim of sustainable development is well represented, but the provisions mainly express the environmental aspects and some economic aspects (such as cooperation and freedom of navigation). Participation of non-state actors is a non-issue. Much will depend on the work and position of the Mekong River Commission and the application of equitable and reasonable utilization. The absence of China and Myanmar hampers an integrated approach. Protocol on Shared Watercourse Systems in the Southern African Development Community, Johannesburg, 18 August 1995. See e.g., Salman (2001a). For the SADC Revised Protocol on Shared Watercourses of 7 August 2000, ILM 20 (2001), 321, see www.sadc.int.

state duty is one of due diligence. Due diligence was defined in the 1872 *Alabama Claims* case: 'a diligence proportioned to the magnitude of the subject and to the dignity and strength of the power which is exercising it.'[45] The standard of conduct required to live up to due diligence will therefore depend on the circumstances of the case such as the magnitude of the threat of harm and the available means to prevent it. In the case of gross neglect by a state, *e.g.* by not adopting necessary measures to protect territory beyond its jurisdiction, or in case of a threat of major harm, this duty is likely to lead to state liability. In its commentary to Article 8 of the Berlin Rules on minimization of environmental harm, the ILA states that the concept of appropriate measures or due diligence at the least includes:

> ...procedural obligations regarding notice and consultation, environmental impact assessment, and a balancing of the social, ecological, and financial costs of an activity, and the ability of the State or States responsible for the activity to bear those costs, as well as the importance of the need the activity is intended to satisfy and the nature and extent of the benefits expected to be realized from the activity.

3.3   International water law

Throughout history, the rise and fall of civilisations has been connected to water availability and management. The importance of water and its regulation has not diminished, given the global water crisis. The international law on water has developed through several attempts to deal with transboundary conflicts.[46] This paragraph examines the evolution of international water law and its major characteristics. The analysis of modern international water law has resulted in the identification of three main principles of international water law: the principle of equitable and reasonable utilization; the no-harm principle; and the principle of cooperation.[47] Equitable and reasonable utilization is discussed in Section 3.3.2. This includes its relation to the no-harm principle, which is further elaborated upon in Chapter 7 since it typically bridges economic and ecological interests. Co-

---

[45] 1872 Geneva Arbitration in the *Alabama Claims* case.

[46] For a thorough and comprehensive analysis of modern international water law see McCaffrey (2001). On modern international water law from the perspective of the Watercourses Convention see Tanzi and Arcari (2001). On international water law see ILA Committee on Water Resources Law (2004), Wouters (ed.) (1997), Caponera (1992), Berber (1959) and www.internationalwaterlaw.org.

[47] *Cf.* the principles included by Tanzi and Arcari (2001) and McCaffrey (2001); see also Hildering (2002) for a comparative review of these books. The no-harm principle is classified as typically combining economic and ecological interests and is therefore dealt with in Section 7.4.2.

operation can be argued to facilitate the first two principles.[48] Cooperation can also be viewed to represent a more general trend toward sustainable development and is therefore discussed in 3.4. Section 3.3.3 makes a preliminary assessment of whether international water law is consistent with the principles of sustainable development.

3.3.1   Evolution of international water law

Water law originated some two thousand years ago.[49] During the centuries, various civilisations have coped with the issue of water allocation and its legal aspects. The rise and fall of early hydraulic civilisations, such as the Egyptian, Mesopotamian, Hindu, Hebrew, and Meso-American civilisations seem to have been closely linked with their development and maintenance of water control systems. During the period in which the Roman Empire flourished (753 BC to the fall of the Western Empire in 476 AD) several water laws were enacted. The compilation of the Eastern Roman Emperor Justinian (527-565 AD) – the *Corpus Iuris Civilis* – contains both classical and post-classical systems of Roman water law.[50] Key principles of law during the Roman period were the distinction between private and public ownership of water (determined by the legal status of land); public supply of water; water use rights; a right to divert water; a right to draw water and a right of access to water; a distinction between uses of water for drinking and domestic purposes, watering of cattle, fishing and transportation, irrigation, industrial purposes, and navigation; prevention of overflow; the prohibition of use of water by the right holder for the sole purpose of damaging his neighbour; and the protection of beneficial uses downstream.[51] Roman law remains influential in Europe as well as other parts of the world. Through European continental law, Roman law was exported with the colonisation of Africa, the Americas, Asia and Australia. European laws influenced or were superimposed on the law of colonies and to this day are reflected in the laws of the former colonies.[52] Other important sources of water law are, *e.g.*, Moslem law, Hindu and Buddhist laws, which have their origin in religious texts.[53]

   Throughout history, laws have developed in response to an experienced need in different regions and throughout various sectors. International regulations on navigation were one of the first important issues of

---

[48] Tanzi (1998), p. 469, concludes that the principle of cooperation will in most cases be the catalyst through which the principle of equity, in conjunction with the no-harm principle, is to operate.

[49] For a detailed history till 1992 see Caponera (1992).

[50] Caponera (1992), p. 41.

[51] See Caponera (1992), pp. 30-43.

[52] British colonies inherited the common law system, while European continental colonies were subjected to variations of Roman law, see Caponera (1992), pp. 75-76 and 80.

[53] See Gupta (2004), pp.12-13, on various layers of water law.

international water law.[54] According to Caflish, freedom of navigation, based on the idea of common interest, was established in Europe in 1815 by the Final Act of the Vienna Congress, peaked around 1920 with the Peace Treaty of Versailles (1919) and the Barcelona Statute on the Regime of Navigable Waterways of International Concern (1921), and declined with fascism as well as during the Cold War and decolonisation. One of the earliest international river institutions is the Central Commission for the Rhine, established in 1831 as a result of a process that can be traced back to 1755, dealing foremost with navigation.[55] According to the UN Food and Agriculture Organization (FAO), between 805 and 1984 more than 3,600 treaties relating to international water resources were negotiated, most of which dealt with aspects of navigation.[56] Since 1814 some have dealt with non-navigational uses as well. Particularly since World War II, issues other than navigation found their way into various bilateral conventions, such as those concerning the use of frontier rivers and the protection of certain rivers or bodies of water from pollution. For a long time, international treaties mainly focused on surface water. The increase in awareness of the biocomplexity of the environment over the last couple of centuries is reflected in treaties, for example, acknowledging the link between surface and groundwater.

Intergovernmental Organizations and Non-Governmental Organizations (NGOs) are of great importance in assisting in the coming into existence of water laws, information gathering and exchange, as a forum for cooperation and in the provision of transparency. Bodies dealing with water include the 23 UN agencies and commissions part of the World Water Assessment Programme such as the FAO and the UN Economic Commissions, the UN ILC, the ILA, IUCN, and numerous water commissions throughout the various continents.[57] The various bodies differ enormously in their scope, constitution, procedures and mandate.

Although international bodies dealing with aspects of water were already developed in the eighteenth century, the increasing involvement of international organizations within international law is a recent trend. The focus on surface water contrasts with the importance of groundwater as a source for human uses, which holds by far the main water reserves, and

---

[54] Caflish (1998), pp. 6-7.

[55] See Caponera (1992), p. 230. See on the international law regime for the River Rhine also Fitzmaurice (2003), pp. 478-482.

[56] See UNFAO (1984), *Systematic Index of International Water Resources by Treaties, Declarations, Acts and Cases, By Basin*, Vol. II, Legis. Study No. 34. According to McCaffrey (2001), p. 59, the first entry into the FAO compilation in 805 is a grant of freedom of navigation on the Rhine by Charlemagne to a monastry.

[57] There is a UN Economic Commission for Asia and the Pacific (ESCAP), for Africa (ECA) for Europe (ECE), for Latin America and the Caribbean (ECLAC) and for Western Asia (ESCWA).

does not necessarily follow the flow of surface waters, even when connected to them.[58]

Important efforts to identify and integrate the established and emerging international water law include the 1966 ILA Helsinki Rules, the 1992 Convention on the Protection and Use of Transboundary Watercourses and International Lakes (ECE Convention), the 1997 Convention on the Law of the Non-Navigational Uses of International Watercourses (Watercourses Convention) and the 2004 ILA Berlin Rules on Water Resources.[59] The Berlin Rules seem to integrate the whole of established and progressively developing international water law and promise to be of great use to modern water management.[60] The Rules deal not only with waters of international drainage basins, but also with waters entirely within a state. They include principles of international law applicable to the management of all waters, internationally shared waters, the rights of persons, protection of the aquatic environments, groundwater, navigation, protection of waters and water installations during war or armed conflict, and state responsibility.

At present, international water law also includes many regional and bilateral treaties that regulate various uses of freshwater resources, both sur-

---

[58] On international law on groundwater, see Salman (1999), Eckstein (1995) and Utton (1982).

[59] The Helsinki Rules on the Uses of the Waters of International Rivers (1966 Helsinki Rules), adopted by the ILA at the fifty-second conference, Helsinki, August 1966, ILA (1966), *Report of the Fifty-Second Conference*, ILA: London. Convention on the Protection and Use of Transboundary Watercourses and International Lakes, Helsinki, 17 March 1992, entry into force: 6 October 1996, 31 *ILM* (1992), 1312, and *UN Doc.* ENVWA-R.53 and Add. 1. Status as of 4 October 2004: 35 parties and 26 signatories. The convention contains 28 articles and 4 annexes (on the definition of the term 'best available technology', on guidelines for developing best environmental practices, on guidelines for developing water-quality objectives and criteria and on arbitration). UN Convention on the Law of the Non-Navigational Uses of International Watercourses, New York, 21 May 1997, *UN Doc.* A/51/869, and 36 *ILM* (1997), 719. The convention contains 37 articles and an annex on arbitration. See for an extensive analysis of the Watercourses Convention Tanzi and Arcari (2001) and McCaffrey (2001). The Convention can, *e.g.*, be found at www.un.org and www.internationalwaterlaw.org. The Berlin Rules on Water Resources (Berlin Rules), adopted at the ILA seventy-first conference, Berlin, August 2004, through Resolution No. 2/2004. The Rules are included in the *Fourth Report of the Committee on Water Resources Law* to the Berlin Conference.

[60] The Rules include references to the Watercourses Convention and the ILA New Delhi Declaration. The ILA Committee on Water Resources Law was re-established in 1991. With the adoption of the Berlin Rules, the Committee on Water Resources Law achieved its mandate and in Resolution No. 2/2004 is recommended to be dissolved. See on the extensive work of the ILA on water, which started in 1954, ILA Committee on Water Resources Law (2004), on www.ila-hq.org the *Sources of the International Law Association Rules on Water Resources* can be consulted for evidence of increasing acceptance in the practice of states of the Berlin Rules, ILA Committee on Water Resources Law (2000) and Bourne (1996).

face and underground.[61] Documents such as the 1992 Rio Declaration and Agenda 21 play an important part in guiding the further development of international water law.

International law regulates water for the following three reasons. Firstly, because of the inter-state character of international river basins. Secondly, because human rights or other principles of international law are affected by water management. Thirdly, in cases and to the extent that the hydrological cycle internationalises national water resources, causing transboundary effects.[62] As stated by Schachter: 'From a scientific standpoint… an international water resource system "includes all the territory within which water occurs or flows as part of a physically interconnected international system."'[63] International law therefore regulates water contained in international rivers such as the contiguous international river the Danube, which is also a successive international river as is the Rhine.[64] It also regulates water in international aquifers such as the Nubian aquifer, international drainage basins such as the Aral Sea Basin or sub-drainage basins, as well as in national rivers, and lakes.[65] Therefore, although many of the principles and most of the treaties referred to in this study address transboundary issues, the research may include freshwater resources not contained by international watercourses shared between two or more countries since it includes water shared through the hydrological cycle and water that is considered an international concern when related to principles of international law such as human rights.

The ECE Convention and the Watercourses Convention are two recent conventions that include almost the whole spectrum of established and emerging international water law. The ECE and Watercourses conventions will now be introduced. The conventions will be further discussed in relation to specific principles of international water law throughout this book. The ECE Convention is discussed before the Watercourses Convention, firstly to emphasise the geographic need for regional solutions and secondly because of its earlier creation and entry into force.

The ECE Convention is an interregional treaty that includes highly developed industrial countries as well as economies in transition, and lower as well as upper riparian states. The ECE Convention came into existence under the auspices of the United Nations Economic Commission for Europe (ECE). The ECE tasks include developing binding international instruments to promote transboundary cooperation and thus to help solve

---

[61] www.transboundarywaters.orst.edu provides an international freshwater treaties database and an international river basins register.
[62] Frant (2003).
[63] Schachter (1977), p. 66, referring to UN Report of Experts, *Management of International Water Resources: Institutional and Legal Aspects*, Doc. ST/ESA/5, 1975, para. 27.
[64] Contiguous international rivers separate two or more states serving as a boundary. Successive international rivers cross successively the territory of two or more states.
[65] The Nubian aquifer is shared by Libya, Egypt, Chad and Sudan. See on the Aral Sea Basin, *e.g.*, Vinogradov (1996).

transboundary problems in the ECE area.[66] The increase in the number of members of the ECE to 55 has given a new urgency to transboundary issues, including issues of freshwater resources.[67] The ECE Convention was adopted at the Resumed 5th Session of the Senior Advisers to the UN ECE Governments on Environmental and Water Problems, held at Helsinki from 17 to 18 March 1992. It entered into force on 6 October 1996, in accordance with Article 26 (1) of the Convention.[68] The ECE Convention is open to member states of the ECE, states having consultative status with the ECE, and to certain regional economic integration organizations constituted by states members of ECE (Article 23).

The ECE Convention contains 28 articles and 4 annexes.[69] The Convention includes provisions dealing with measures to be taken by all parties (Part I) such as on monitoring the conditions of transboundary waters (Article 4), provisions only relating to riparian states (Part II) such as on cooperation through arrangements (Article 9), and institutional and final provisions (Part III). Article 1 of the ECE Convention defines transboundary waters to include both surface water and groundwater involving two or more states.[70] All parties to the ECE Convention are obliged to prevent,

---

[66] Bosnjakovic (1998), p. 49.

[67] Between 1990 and 1995 the membership increased from 35 to 55 countries, including 27 countries in transition. Bosnjakovic (1998), p. 47.

[68] As of 4 October 2004, there were 35 parties (Albania, Austria, Azerbaijan, Belarus, Belgium, Bulgaria, Croatia, Czech Republic, Denmark, Estonia, Finland, France, Germany, Greece, Hungary, Italy, Kazakhstan, Latvia, Liechtenstein, Lithuania, Luxembourg, Netherlands, Norway, Poland, Portugal, Republic of Moldova, Romania, Russian Federation, Slovakia, Slovenia, Spain, Sweden, Switzerland, Ukraine and the European Community) and 26 signatories to the ECE Convention. See www.unece.org/env/water/welcome.html.

[69] Article 1 of the ECE Convention defines transboundary waters, transboundary impact, party, riparian parties, joint body, hazardous substances, and best available technology. Article 2 contains general provisions. Article 3 deals with prevention, control and reduction. Article 4 is on monitoring the conditions of transboundary waters. Articles 5, 6, 7 and 8 relatively concern research and development, exchange of information, responsibility and liability, and protection of information. In Part II, the obligation to enter into bilateral or multilateral agreements is arranged for in article 9 (equality and reciprocity, catchment area or part(s) thereof, joint bodies). Article 10 deals with consultations, article 11 with joint monitoring and assessment, article 12 with common research and development, article 13 with exchange of information between riparian parties, article 14 with warning and alarm systems, article 15 with mutual assistance, and article 16 with public information. Annex I further defines the term 'best available technology', Annex II contains guidelines for developing best environmental practices, Annex III deals with guidelines for developing water-quality objectives and criteria, and Annex IV is on arbitration. See on a comparison of water management between the EU/ECE and SADC regions, including on the 1997 SADC-EU Maseru Conference, Savenije and Van der Zaag (2000a) and (2000b), Van der Zaag and Savenije (1999) and (2000).

[70] Article 1(1) ECE convention: '"Transboundary waters" means any surface or groundwaters which mark, cross or are located on boundaries between two or more

control and reduce any transboundary impact (Article 2.1), relating to water and pollution and use. Article 2 includes reference to: reasonable and equitable use of transboundary waters, equality and reciprocity, and catchment areas. It also includes principles such as the precautionary principle, polluter-pays principle, sustainable development in water resources ('water resources shall be managed so that the needs of the present generation are met without compromising the ability of future generations to meet their own needs'). Article 3 elaborates upon the measures to be taken to prevent, control and reduce transboundary impact, including emission of pollutants and environmental impact assessment.

In contrast, the 1997 Watercourses Convention has universal aspirations and is open to all states and regional economic integration organizations (Article 2). In 1971, the UN International Law Commission (ILC) began the study of the law of non-navigational uses of international watercourses.[71] In 1991, the ILC presented draft articles to the United Nations General Assembly (UNGA) on the non-navigational uses of international watercourses. The UNGA adopted a modified version of these articles in 1997, *i.e.* the Watercourses Convention.[72] Although a convincing majority of states voted in favour of the adoption of the Watercourses Convention, the slow rate of ratification at present indicates that it is doubtful whether the Convention will enter into force.[73] The fact that it took the Watercourses Convention more than twenty years to come into existence illustrates both the differences of opinion between states on the existing and emerging law of non-navigational uses of international watercourses and the difficulties in formulating a worthwhile common denominator. Despite the adoption of the Watercourses Convention discussion on international

---

States; wherever transboundary waters flow directly into the sea, these transboundary waters end at a straight line across their respective mouths between points on the low-water line of their banks'.

[71] The ILC was established in 1947 as a permanent subsidiary body by the UNGA, based on Article 13.1(a) of the UN Charter which provides that the General Assembly shall initiate studies and make recommendations for the purpose of encouraging the progressive development of international law and its codification. The work of the ILC further encompasses preliminary drafts of the UNCLOS and Draft Articles on State Responsibility.

[72] The Watercourses Convention was adopted by the UN General Assembly on its 51st session by resolution A/RES/51/229 of 21 May 1997.

[73] The Convention could not be adopted by consensus but had to be put to vote: 103 votes in favour, 3 against (Turkey, China, and Burundi) and 27 abstentions. Under Article 36 of the Watercourses Convention, 35 instruments of ratification, acceptance, approval or accession are needed for the convention to enter into force. As of 4 October 2004, there were 12 parties (Finland, Hungary, Iraq, Jordan, Lebanon, Namibia, Netherlands, Norway, Quatar, South Africa, Sweden and Syrian Arab Republic) and 16 signatories to the Watercourses Convention.

water law continues. The response of different authors to the Watercourses convention varies from scepticism to quite optimism.[74]

The scope of the Watercourses Convention (Article 1) concerns non-navigational uses of international watercourses and their waters and measures of protection, preservation and management related to it.[75] Although navigation is adequately regulated under international law, it is included, according to Article 1 of the Convention, to the extent that it interacts with non-navigational uses, thereby enhancing the potential for an integrated approach. In Article 2(a), a watercourse is defined as follows: 'a system of surface waters and groundwaters constituting by virtue of their physical relationship a unitary whole and normally flowing into a common terminus'. The definition thus has a much wider range than traditional definitions of watercourses. However, confined groundwater is excluded by the definition, since confined water does not interact with surface water within the limits of the time frame under consideration.[76] The international character of a watercourse is determined by its traversing more than one state.

One can critique the supposed framework character of the Watercourses Convention, since it does not provide minimum standards to be further developed by the parties and the parties are free to deviate from the provisions by agreement.[77] The ambiguity of its substantive principles is another source of criticism.[78] Especially from the perspective of international environmental law, the provisions of the Watercourses Convention can be argued not to be progressive enough.[79]

On the other hand, it can be argued that the Watercourses Convention will have influence, even if not in force, since many provisions are a codification of customary international law and it furthermore contributes to the progressive development of international law.[80] Moreover, the adoption by the UNGA of the Watercourses Convention may be said to indicate a certain degree of consent on its content. Furthermore, the Convention, even though not in force, has been referred to by the International Court of Justice (ICJ) in the 1997 *Gabcíkovo-Nagymaros* judgement (Hun-

---

[74] Opinions expressed by authors on the ILC Draft Articles are likely to apply similarly to the provisions of the Watercourses Convention since the differences are marginal.

[75] The Watercourses Convention, according to its Article 1: 'applies to uses of international watercourses and of their waters for purposes other than navigation and to measures of protection, preservation and management related to the uses of those watercourses and their waters.'

[76] On confined groundwater, the ILC adopted a resolution encouraging states to apply the provisions of the Watercourses Convention, Report of the International Law Commission on the Work of its Forty-Sixth Session, *UN Doc.* A/49/10 (1994), at 326.

[77] Hey (1998).

[78] Nollkaemper (1996) and Schwabach (1998).

[79] Brunnée and Toope (1997).

[80] Tanzi and Arcari (2001) and McCaffrey (2001).

gary/Slovakia). Its contents are also included in, for example, the 2000
SADC Revised Protocol.[81]

In a sense, the weakness of the Watercourses Convention is its
strength: its flexibility and the possibility to deviate from it enables integrated water management in all sorts of historical, hydrological and geographical situations and allows for the further evolution of international
water law.

3.3.2   Equitable and reasonable utilization

The allocation and use of water between states is mainly regulated by the
principle of equitable and reasonable utilization.[82] The equitable and reasonable utilization of watercourses by states can be successfully argued to
constitute the main principle of international water law.[83] This principle
has been codified in Article 5 of the Watercourses Convention, which
states:

> 1.   Watercourse States shall in their respective territories utilize an in-
> ternational watercourse in an equitable and reasonable manner. In par-
> ticular, an international watercourse shall be used and developed by
> watercourse States with a view to attaining optimal and sustainable
> utilization thereof and benefits therefrom, taking into account the in-
> terests of the watercourse States concerned, consistent with adequate
> protection of the watercourse.
> 2.   Watercourse States shall participate in the use, development and
> protection of an international watercourse in an equitable and reason-
> able manner. Such participation includes both the right to utilize the
> watercourse and the duty to cooperate in the protection and develop-
> ment thereof, as provided in the present Convention.

Besides sustainable utilization, this Article refers to optimal utilization.[84]
Optimal utilization, for example, stimulates fishing to the point that the
species of fish is most reproductive in order to maintain its population,
which may be economically efficient but hardly takes other effects such as
on the well-being of the fish and on the larger ecosystem into account. It

---

[81] Gupta (2004), p. 27.
[82] For a thorough elaboration of equitable utilization of water resources, see *e.g.*, Kaya
(2003) and Lipper (1967).
[83] See McCaffrey (2001), p. 345: 'Equitable utilization is the fundamental rule governing the use of international watercourses.' According to Caflisch (1998), p. 13: 'This
principle, which today governs the attribution of shared water resources, has its roots in
the judicial practices of federal States such as the United States, Germany and Switzerland (...).' In Article 12 of the Berlin Rules, equitable utilization concerns waters of an
international drainage basin and is not limited to watercourses.
[84] See Benvenisti (2002).

moreover contains a right of states to use water but limits the ways in which to use it and requires cooperation.

Article 6 of the Watercourses Convention further substantiates Article 5 by elaborating on the relevant factors and circumstances to be taken into consideration in obtaining reasonable and equitable utilization. The list of factors is not exclusive. It refers to: natural characteristics; social and economic needs of the watercourse states concerned; the population dependent on the watercourse in each watercourse state; the effects of the use or uses of a watercourse in one watercourse state on other watercourse states; existing and the potential uses of the watercourse; conservation, protection, development and economy of the use and the costs of related measures; and alternatives.[85] These elements include emphasis on economic uses of water resources and probably require cost-benefit analyses. According to Article 6.2, when consultations are needed they are to take place in a spirit of cooperation. Article 6.3 of the Watercourses Convention requires that in allocating fresh water between its different uses all relevant factors and circumstances have to be taken into account and weighed against each other and a conclusion is to be reached on the basis of the whole.[86]

A further qualification of the entitlement of states to utilize a watercourse is formed by the obligation not to cause harm beyond national jurisdictions as formulated in Article 7.[87] Equitable and reasonable utilization and the no-harm principle are often viewed as conflicting and competing principles.[88] The fact that at present practically any use of fresh water might lead to harm because of over-exploitation of fresh water is often referred to as the main reason for considering the principle of equitable and reasonable use as more suitable for regulating water allocation.[89] Another argument for preferring equitable and reasonable utilization could be that the no-harm principle can be regarded primarily as a principle of demarcation rather than of cooperation. On the other hand, considering the continuing degradation of the environment, the prohibition of certain harm would seem all the more urgently needed. Moreover, cooperation, including prompt exchange of information facilitated by joint bodies, is

---

[85] See Section 3.3.3 on the relevance of the inclusion of future uses.

[86] The concept of a balance of interests was already recognised in the *Diversion of Water from the Meuse* Case and the *Helmand River Delta* Case.

[87] See Section 7.4.2 on the no-harm principle, categorised as the principle that in particular combines economic and ecological interests.

[88] In the coming into existence of the Watercourses Convention, the prevalence of either the no-harm principle or equitable utilization over the other gave rise to many discussions. Finally the "package deal" represented by Articles 5 to 7 was accepted by 38 votes to 4 (China, France, Tanzania, Turkey) and 22 abstentions. The alleged conflict between the no-harm principle and equitable and reasonable utilization, however, remains an issue of debate. See, *e.g.*, ILC Summary Records of the Meetings of the Forty-sixth Session, 2336th Mtg., *Yearbook of International Law*, 1 (1994), p.167, Tanzi and Arcari (2001), pp. 175-179, and Caflisch (1998), pp. 12-16.

[89] Caflisch (1998), p. 12.

equally a condition necessary for implementing the no-harm principle. Nowadays, it is increasingly argued that the principle of equitable and reasonable utilization and the no-harm principle supplement one another.[90] It is interesting to note that in the Berlin Rules Article 12 on equitable utilization includes due regard for the no-harm principle and Article 16 on avoidance of transboundary harm includes due regard for the right of a basin state to make equitable and reasonable use of waters.

In the *Gabcíkovo-Nagymaros* case, the ICJ affirmed the right to an equitable and reasonable share of a watercourse.[91] Paragraph 78 of the ICJ Judgment states:

> The suspension and withdrawal of that consent constituted a violation of Hungary's legal obligations, demonstrating, as it did, the refusal by Hungary of joint operation; but that cannot mean that Hungary forfeited its basic right to an equitable and reasonable sharing of the resources of an international watercourse.

In depriving Hungary of its right to an equitable and reasonable share of the natural resources of the Danube, the proportionality required by international law was not respected.

The principle of equitable and reasonable utilization has also been expressed in soft law instruments, such as Article 4 of the ILA Helsinki Rules: 'each basin State is entitled, within its territory, to a reasonable and equitable share in the beneficial uses of the waters of an international drainage basin.'[92]

The principle of equitable and reasonable utilization provides states not only with a right to use water but also qualifies state sovereignty. In the application of equitable and reasonable utilization, the factors to be taken into account are virtually unlimited. Depending on such factors and their given weight, the principle could play an important part in the achievement of sustainable development.

---

[90] Tanzi and Arcari (2001), p. 302, in analysing the Watercourses Convention, state that the equitable utilisation principle and the no harm rule are part and parcel of the same normative setting. See also McCaffrey (2001), p. 380: 'Far from being incompatible with equitable utilization, therefore, the no-harm obligation is a necessary and integral part of the equitable utilization process.'

[91] In the *Gabcíkovo-Nagymaros* case, both the *River Oder* case and the *Diversion of the Waters of the Meuse* case of the PCIJ were relied on: *Territorial Jurisdiction of the International Commission of the River Oder*, 1929, PCIJ Ser. A No. 23 at 5; *Diversion of Water from the Meuse*, 1937, PCIJ Ser. A/B No. 70 at 4. The PCIJ *Oscar Chinn* case, 1934, PCIJ Ser. A/B No. 63 at 65, relates to international water law as well.

[92] Factors to be taken into account are included in Article 5.

3.3.3 Sustainable development in international water law

In the *Gabcíkovo-Nagymaros* case on the Danube river, the main ICJ case on non-navigational uses of freshwater resources, the ICJ acknowledged the relevance of sustainable development in international law. This section now reviews the extent to which sustainable development is actually part of international water law as reflected in the ECE and Watercourses conventions and the principle of equitable and reasonable utilization.

Both the Watercourses Convention and the ECE Convention refer to sustainable development. Article 2.5(c) of the ECE Convention states that parties, when taking measures to prevent, control and reduce any transboundary impact, shall be guided by the principle that: 'Water resources shall be managed so that the needs of the present generation are met without compromising the ability of future generations to meet their own needs'. In the preamble of the Watercourses Convention the conviction is expressed 'that the framework convention will ensure the utilization, development, conservation, management and protection of international watercourses and the promotion of the optimal and sustainable utilization thereof for present and future generations'. Parties are therefore to interpret the Watercourses Convention in line with sustainable development. Article 5 of the Watercourses Convention obliges watercourse states to use and develop watercourses with a view to attaining optimal and sustainable utilization. Article 24.2(a) of the Watercourses Convention is the only article that includes the term sustainable development in stating that management refers to: 'Planning the sustainable development of an international watercourse'.[93]

Both the ECE and Watercourses conventions address environmental concerns in their provisions, further discussed in later chapters. The ECE Convention does not refer to developmental concerns as such. This may be partly explained by its geographical range, although it includes economies in transition. In the preamble of the Watercourses Convention, the situation of developing countries is emphasised, recalling the principles and recommendations adopted by the UNCED 1992 in the Rio Declaration and Agenda 21. In establishing the elements to be taken into account, both the social and economic needs of watercourse states as well as the population dependent on the watercourses are referred to (Article 6.1(b) and (c)). The inclusion of potential uses in Article 6.1(e) of the Watercourses Convention, rather than mentioning only existing uses, provides increased pos-

---

[93] The 1994 Draft Articles of the Watercourses Convention did not include specific reference to sustainability in the articles on equitable and reasonable utilization. For a discussion of sustainable development in the drafting history of the Watercourses Convention, see Fuentes (1999), pp. 120-122.

sibilities for the developmental side of sustainable development.[94] Therefore, prior appropriation – under international law reflected in historical
rights – is not granted inherent priority. According to the ILC the reference
to both existing and potential uses in the Article is: 'in order to emphasise
that neither is given priority, while recognising that one or both factors
may be relevant in a given case.'[95] The beneficial or adverse consequences
of the existing and future uses are to be taken into account.[96] Principles
such as that of prior appropriation can therefore under circumstances be
set aside. The main opportunity for developmental interests to be taken
into account would seem to lie within the application of the principle of
equitable and reasonable utilization.

The principle of equitable and reasonable utilization shares characteristics with sustainable development in several ways.[97] According to Wouters
and Rieu-Clarke: 'This principle provides, indeed requires, that States take
into consideration the factors tied to sustainable development of the resource, thus providing the legal framework for operationalising this concept.'[98] The principle articulates a process that integrates the many aspects
of water allocation on a case by case basis. Equitable and reasonable utilization is therefore a step forward toward an integrated approach, while at
the same time acknowledging the uniqueness of each case. By weighing all
the interests involved, it takes into account the economic, social and ecological factors as a whole and aims for a balance. In the Watercourses Convention, the principle of equitable and reasonable utilization is furthermore
placed in a wider framework that requires it to take into account sustainable development and several of its elements.[99] The principle can therefore
be instrumental for international law to encourage fresh water to be allocated in a sustainable manner.

However, certain characteristics of the principle create uncertainty
over its compatibility with sustainable development. A difference between
the principle of equitable and reasonable utilization and sustainable development, lies within the reasons that brought them into being. Equitable

---

[94] On the inclusion of future uses, see Tanzi and Arcari (2001), pp. 132-134, who also
refer to the continuing division in the legal literature on the pros and cons of the protection of such uses.

[95] 1994 Report of the International Law Commission on the Work of its Forty-Sixth
Session, *UN Doc.* A/49/10, p. 233.

[96] See Tanzi and Arcari (2001), pp. 133-134, referring to the Report to the UNGA of
the Working Group of the Whole.

[97] For a thorough analysis of sustainable development and equitable utilization of international watercourses see Fuentes (1999).

[98] Wouters and Rieu-Clarke (2001), p. 3.

[99] See also Section 3.3.3. The term 'sustainable development' was not incorporated in
the principle of equitable and reasonable utilization itself in the Watercourses Convention. The Dutch proposed to include sustainable development in Article 5 of the Watercourses Convention on equitable and reasonable utilization and the Finish proposed to
include it in Article 6.1, see *ILC Doc.* A/C.6/51/NUW/WG/CRP.18.

and reasonable utilization was not created as a response to those problems that call for sustainable development. Rather, its raison d'être was the resolution of conflicts between riparian states.[100] Equitable and reasonable utilization in itself therefore does not require states to aim for the common goal of sustainable development. Sustainability can be compromised. For example, two states might end a conflict by dividing waters of a shared water resource beyond the replenishment rate.[101] The process of equitable and reasonable utilization does not seem to take into account the interrelationship of water and ecosystems. In order to contribute to the achievement of sustainable development, the principle is moreover to extent to non-riparian states and non-state actor participation.

It can also be argued that the principle of equitable and reasonable utilization is somewhat obscure. The elements that must be taken into account are practically unlimited and the weight attributed to each is left uncertain, with no inherent priority of one use over another.[102] The outcome of equitable and reasonable utilization on a case by case basis is therefore highly unpredictable. Another factor contributing to this lack of clarity, is the somewhat ambiguous nature of principles and their interrelationship within the Watercourses Convention, requiring more substantive principles to provide better guidance and identification of the content of the principle.[103] According to Nollkaemper: 'The agenda for the future discourse of water law has already been set and there is little doubt that that will converge around the notion of protection of vital human needs, ecosystem protection and sustainability.'[104] Vital human needs, ecosystem protection and sustainability are explicitly cited in the Watercourses Convention but the uncertain outcome of equitable and reasonable utilization leaves the protection of those issues at the mercy of the watercourse states concerned.

3.4 Trends in favour of sustainable development

Within international water law trends can be identified that are likely to contribute to the achievement of sustainable development. The most im-

---

[100] Fuentes (1999), p. 200.

[101] Section 6.2.2 elaborates on the replenishment rate.

[102] Nollkaemper (1996) states that the principle is highly indeterminate and relies on a contextual balance of all relevant factors and circumstances and each interest can be overridden by another. The author goes even so far as to say that 'the principle is little more than an open-ended framework for political compromise without an independent legal identity.'

[103] See, e.g., Tanzi and Arcari (2001), pp. 95-96: 'Although the status of the equitable utilisation principle as a 'cornerstone' of the general law of international watercourses is frequently postulated both in theory and in practice, some uncertainty remains as to its normative impact, if not its actual content.'

[104] Nollkaemper (1996), p. 53.

portant changes of general direction are those toward an integrated approach, those reflecting increased cooperation and those that underline community interests. These developments are now further discussed.

3.4.1   From fragmentation toward integration

The development of international law on water relating to sectors such as navigation or industry on the one hand and uses such as domestic or ecological on the other has left us with a fragmented law.[105] Like international water law, international law itself used to consider the various fields of sustainable development separately. The global problems of environment and development, including problems in relation to water, call for an integrated response. An integrated approach requires coordination of processes, a comprehensive consideration of related factors and various kinds of cooperation such as exchange of information. Although it can be argued that incorporating too much within the concept of sustainable development will only reduce its efficacy, sustainable development cannot possibly be achieved if an integrated approach to social, economic and ecological factors is not adopted. Denial of the interrelationship between the different elements might simplify matters in theory but in practice would amount to a refusal to face the nature of the problems that have given rise to the demand for sustainable development in the first place. No matter how convenient, managing the water levels of urban and nature areas without acknowledging the interaction will not stop water in finding its way from the high water level of a wetland to the low water level of a neighbouring city, possibly causing damage to both. Nor would it be efficient to deny the interaction of ecosystems worldwide.

The growing awareness of the interrelationship between freshwater resources and the emerging international law on sustainable development is increasingly reflected in international water law. A trend away from fragmentation and toward integration within international water law is reflected in the much more integrated approach taken by the ECE and Watercourses conventions. The ECE Convention and the Watercourses Convention constitute a major advance over most of the earlier agreements and international law on fresh water in that they have abandoned the sectoral approach and the division between surface and groundwater. Several of their Articles will be further discussed throughout this study.

Turning to soft law, the Rio Declaration can be regarded as an attempt to achieve a more integrated approach. The need to take a balanced and integrated approach to environment and development questions is further underlined by Agenda 21.[106] Chapter 18 of Agenda 21 emphasises the essential nature of freshwater resources for hydrosphere and ecosystems and acknowledges their part in the hydrological cycle, the influence of climate

---

[105] See Chapter 2 of this book on water uses and sectors.
[106] Preamble 1.2 of Chapter 1 of Agenda 21.

change, atmospheric pollution and sea-level rise, and thereby its biocomplexity. Chapter 18 moreover states that integrated water resources planning and management are needed and that the many interests involved in utilization of water resources must be recognised.[107] It furthermore calls for holistic management and an integrated approach at the level of the catchment basin or sub-basin.[108] According to the preamble of the ILA New Delhi Declaration, sustainable development is now widely accepted as a global objective and is a matter of common concern, to 'be integrated into all relevant fields of policy in order to realize the goals of environmental protection, development and respect for human rights'.

In the ILA New Delhi Declaration, a principle of integration and interrelationship, in particular in relation to human rights and social, economic and environmental objectives, is included (Principle 7). A principle of integration and interrelationship is not established in general international law but might be evolving considering the increasing importance and acceptance of its ingredients: the relation between social, economic and ecological aspects of principles of international law relating to sustainable development and between the needs of present and future generations (cf. Principle 7.1); the need for an integrated approach taken at all levels of governance and by all sectors of society (cf. Principle 7.2); the aim to overcome conflicts of social, economic and ecological interests between states, facilitated by institutions (cf. Principle 7.3); and the need to interpret and apply the applicable principles of international law taking into account their interrelationship (cf. Principle 7.4).

The Draft International Covenant on Environment and Development of the World Conservation Union (IUCN) articulates an integrated approach in Article 19 on water:[109]

> The Parties shall take all appropriate measures to maintain and restore the quality of water, including atmospheric, marine, ground and surface fresh water, to meet basic human needs and as an essential component of aquatic systems. The Parties also shall take all appropriate measures, in particular through conservation and management of water resources, to ensure the availability of a sufficient quantity of water to satisfy basic human needs and to maintain aquatic systems.

Although international water law is on its way, it has not yet evolved to the point where it deals comprehensively with integration. For example, waters not connected to a "common terminus" as formulated in international water law are still mostly omitted. The exclusion of so-called confined groundwater from the Watercourses Convention, including them in a

---

[107] Chapter 18.1 and 18.3 of Agenda 21.
[108] A further elaboration upon a catchment basin approach is contained in Section 6.4.1.
[109] See IUCN (2004), p. 71. On the 1995 IUCN Draft Covenant on Environment and Development see Boyle (1999), p. 71.

resolution instead, can result in the exclusion of important waters such as involved in disputes in the Middle East.[110] Moreover, the information gap, especially in the field of groundwater, poses serious constraints on the possibilities of operationalising an integrated approach to water management.[111] Chapter VIII on groundwater of the 2004 Berlin Rules does apply to all aquifers, including those not connected to surface water and those that receive no significant contemporary recharge. It also includes a duty to acquire information (Article 39).

3.4.2   From delimitation toward cooperation

Another trend that seems to contribute to achieving sustainable development is that toward cooperation. Cooperation relates to all national, regional and international levels and can be between states as well as with other parties. There are miscellaneous forms of cooperation, varying from friendly relationships, voluntary exchange of information, joint monitoring or research to duties such as to exchange specific information, notification, early warning, entry into consultations or agreements, or establishment of joint bodies for joint implementation. Cooperation is moreover required within the context of the obligation of peaceful settlement of disputes.

International law used to focus on territorial delimitation, as reflected mainly in the strong position of the principle of sovereignty of states.[112] Delimitation was the main focus of international water law as well. An early field of regulation of water resources, apart from navigation, concerns the establishment of boundaries. Political boundaries fixed by international watercourses are at present arranged by treaties.[113] The transboundary impact of many freshwater uses and increased water stress in many regions, render delimitation rights and duties accompanying transboundary water resources inadequate for the solution of present-day problems. It can moreover be concluded from Chapter 2 that the problems relating to water uses require cooperation as a condition to balance the interests. International cooperation is also required to achieve the goal of access to water for all. These issues call for cooperation at all levels to facilitate the implementation of sustainable development in freshwater resources management. The need for cooperation does not exclude delimitation. It could even be argued that fruitful cooperation can only take place when

---

[110] According to Scanlan (1996), p. 2229: 'Essentially, if Middle Eastern states fail to acknowledge the resolution on confined groundwater, the Draft Articles [of the Watercourses Convention], after more than twenty-five years spent researching the subject, would be rendered worthless for the purpose they were written.'

[111] Gleick (2000), pp. 40-42.

[112] Green Cross International (2000).

[113] Caflisch (1998), p. 5: 'Boundaries in waterways are almost invariably drawn by treaty, so that the real problem will be one of interpreting the provisions of the treaty.' Apart from the legal position, potential problems lay in the actual interpretation on the ground.

delimitation is clear, for example, when the parties and their jurisdiction can be identified.

Articles 1 and 2 of the UN Charter not only underline the importance of international peace and security, but also of such principles as international cooperation, harmonisation and peaceful settlement of disputes.[114] Cooperation is regarded not only as a principle but also as a requirement of modern international law.[115]

Principles of cooperation are well presented in both the ECE and Watercourses conventions. Under Article 6, the parties to the ECE Convention are under obligation to provide for the widest exchange of information. Article 9 of the ECE Convention obliges the parties to enter into bilateral or multilateral agreements, on the basis of equality and reciprocity, embracing all relevant issues covered by the Convention, and if necessary to adapt existing arrangements in order to eliminate contradictions with the basic principles of the ECE Convention. The agreements have to establish joint bodies whose tasks are set out in Article 9. Under the ECE Convention, riparian parties are obliged to enter into consultations (Article 10). Furthermore, riparian parties are obliged to establish joint monitoring and assessment (Article 11), to undertake common research and development (Article 12), to exchange information between riparian parties (Article 13), to notify any critical situations and provide mutual assistance (Article 14), and to make information available to the public (Article 16).

The Watercourses Convention in its preamble explicitly affirms the importance of international cooperation and good neighbourliness. The general obligation to cooperate is laid down in Article 8 of the Watercourses Convention, in which the establishment of joint mechanisms or commissions is suggested.[116] Article 8 of the Watercourses Convention states:

> 1. Watercourse States shall cooperate on the basis of sovereign equality, territorial integrity, mutual benefit and good faith in order to attain optimal utilization and adequate protection of an international watercourse.
> 2. In determining the manner of such cooperation, watercourse States may consider the establishment of joint mechanisms or commissions, as

---

[114] Armed conflict is commonly regarded as inherently destructive of sustainable development. See for example Rio Declaration Principles 24 and 25.

[115] See Perrez (2000), pp. 330-331, for the need of cooperation in modern international law and the concept of cooperative sovereignty. See also ILA Committee on Legal Aspects of Sustainable Development (2002), p. 7, discussing the duty to cooperate towards global sustainable development and protection of the global environment.

[116] According to the commentary to Article 11 of the Berlin Rules on cooperation: 'The duty of cooperation is the most basic principle underlying international water law.', ILA Committee on Water Resources Law (2004), p. 20. On the prospects of a treaty on the Euphrates and Tigris between Turkey, who rejected the Watercourses Convention, Syria and Iraq after the adoption of the Watercourses Convention, see Lien (1998).

deemed necessary by them, to facilitate cooperation on relevant measures and procedures in the light of experience gained through cooperation in existing joint mechanisms and commissions in various regions.

A duty to exchange data and information can be found in Article 9 of
the Watercourses Convention.[117] This exchange of information is further
identified for planned measures in Articles 11-19 of the Watercourses
Convention, which mention consultation, negotiation and regulating notification. Many other provisions of the Watercourses Convention also contain principles of cooperation. In preventing, controlling and reducing
transboundary harm, the riparian parties must cooperate on the basis of
equality and reciprocity (Article 2.6). Under Article 3 of the Watercourses
Convention, parties may consider harmonising other agreements with the
basic principles of the Watercourses Convention, although allowing existing or future agreements to deviate from the convention. Article 4 of the
Watercourses Convention states the right of watercourse states to take part
in a watercourse agreement that applies to the total international watercourse, to participate in consultations concerning watercourse agreements
and to agreements covering part of the watercourse if the implementation
affects it. According to Article 5 of the Watercourses Convention, watercourse states have the duty to cooperate in the protection and development
of the watercourse. When needed in the application of equitable and reasonable utilization, the Watercourses Convention requires watercourse
states to enter into consultations in a spirit of cooperation (Article 6). Before the implementation of any planned measures that may have adverse
effects, timely notification is required by the Watercourses Convention,
accompanied by data and information, including those resulting from environmental impact assessments (Article 12). Articles 13 to 19 set out the
procedural arrangements in the case of such (absence of) notification.
Other references to cooperation in the Watercourses Convention include
Article 25, on the regulation of the flow of an international watercourse,
and Article 28 which applies to emergency situations where the prompt
notification of other potentially affected states and competent international
organizations is required. The duty to cooperate is emphasised by many
regional conventions as well, such as the 1995 Mekong Agreement.[118]
    Under customary international law there seems to be a duty of prior
consultation. The 1957 *Lac Lanoux* arbitration (Spain v. France) affirms
that prior consultation and negotiation constitute a principle of customary

---

[117] According to the commentary to Article 18 of the Berlin Rules, within the chapter
on rights of persons, a right to information is well-established, although its precise
contours can be debated. An interesting movement to enable exchange of information
without the obstacle that patenting can pose to, *e.g.*, the access of developing countries
to information, is the 'open source' movement, entailing a whole body of affordable
information technology, see www.opensource.org.
[118] See Section 3.2.3, note 44, on the Mekong Agreement.

law.[119] In the case that a use of shared resources may involve serious injury to the rights or interests of another state, it is a duty under international law to give prior notice, consult and negotiate.[120]

In the *Lac Lanoux* case, the Arbitral Tribunal stated that, although the use of a shared water resource does not require agreement, such agreement should be striven for and under the rules of good faith the upstream state should take into account the various interests involved. In the arbitration, a right to information was established.[121] In its judgement on the *Gabcíkovo-Nagymaros* case, the ICJ confirmed the importance of cooperation in the use of shared water resources.

The need for cooperation is also expressed in many of the additional sources of international law. Under Article XXIX of the 1966 ILA Helsinki Rules, it is recommended that basin states provide all relevant and reasonably available information relating to the waters of a drainage basin within its territory to other basin states.[122] Another example is the UNGA Declaration on Principles of International Law concerning Friendly Relations and Cooperation Among States.[123] The Rio Declaration requires states to cooperate for reasons such as to strengthen capacity-building and to promote a supportive and open international economic system.[124] Principle 18 of the Rio Declaration deals with the obligation of states to notify other states in case of emergencies.[125] Emergency situations include disasters such as those caused by floods and heavy pollution. Principle 19 of the Rio Declaration reflects the duty of potentially affected states to notify and consult on activities that might have significant adverse transboundary environmental effect. Agenda 21 underlines the need for a global partnership for sustainable development.[126] Objectives formulated in the Plan of Implementation of the WSSD, include: 'Strengthening international cooperation aimed at reinforcing the implementation of Agenda 21 and the outcomes of the Summit.'[127]

Greater cooperation can go a long way toward anticipating controversies and preventing conflicts.[128] International water resources as a reason

---

[119] *Lac Lanoux* Arbitration (Spain v. France), award of 1957, 24 *ILR* 101, on the utilization by France of the Lake Lanoux in the Pyrenees, shared with Spain.

[120] *Cf.* Birnie and Boyle (2002), p. 319.

[121] *Lac Lanoux* Arbitration, 119: 'A state wishing to do that which will affect an international watercourse cannot decide whether another state's interests will be affected; the other state is sole judge of that and has the right to information on the proposals.'

[122] ILA Helsinki Rules.

[123] UN General Assembly Declaration on Principles of International Law concerning Friendly Relations and Cooperation Among States in Accordance with the Charter of the United Nations, 1970, UNGA Res. 2625 (XXV).

[124] Rio Declaration Principles 9 and 12.

[125] On emergency cooperation, see Birnie and Boyle (2002), pp. 322-323.

[126] Preamble 1.1 of Chapter 1 of Agenda 21.

[127] WSSD Plan of Implementation, para. 121(i).

[128] Benvenisti (1996), p. 415: 'Just as the prospect of water scarcity facing ancient communities moved them to establish reliable collective-action mechanisms, so the

for cooperation by far outweighs water as a reason for conflict between states. According to the WWDR:[129]

> The last fifty years have seen only thirty-seven acute disputes (those involving violence) while, during the same period, approximately 200 treaties were negotiated and signed. The total number of water-related events between nations, of any magnitude, are likewise weighted towards cooperation: 507 conflict-related events, versus 1,228 cooperative ones, implying that violence over water is not strategically rational, effective or economically viable.

In cases where disputes nevertheless arise, cooperation is required to fulfil the principle of peaceful settlement of conflicts.[130] Where disputes occur, peaceful settlement is the prescribed way of resolving them.[131] Article 33 of the UN Charter mentions negotiations, enquiry, mediation, conciliation, arbitration and judicial settlement, resort to regional agencies or arrangements, or other peaceful means of the parties' choice, as means of settling a dispute. Apart from delimitation cases to establish boundaries, a surprisingly small number of disputes over freshwater uses have been referred to the ICJ, namely the *Gabcíkovo-Nagymaros* case (Hungary/Slovakia), the *Kasikili/Sedudu* dispute (Namibia/Botswana), and a dispute pending on the *Niger river* (Niger/Benin). Settling disputes over freshwater resources is nevertheless often provided for in treaties between states. The International Tribunal for the Law of the Sea (ITLOS) has also dealt with interesting cases with an impact on principles governing freshwater management, such as the *Southern Bluefin Tuna* cases and the case concerning *Land Reclamation by Singapore in and around the Straits of Johor.*[132] Apart from the hy-

---

current growing demands for water, which are expected to intensify in the twenty-first century, can encourage cooperation on a regional basis, and not necessarily lead to water wars.' On the role of institutions in prevention and resolution of conflicts over shared river basins in Africa, see Okaru-Bisant (1998). See Salman and Uprety (2002) on cooperation and conflicts over rivers in South Asia, and Fitzmaurice (2001), pp. 428-467, on water resource cooperation illustrated by the effective cooperation in the Nordic States and the potential conflict over the Jordan River. On Northern Europe, see also Fitzmaurice and Elias (2004).

[129] WWAP (2003), p. 312. However, water does tend to get used as an instrument in conflicts between states, see Section 5.3.3. Furthermore, conflicts over water often occur at the national and community levels.

[130] www.transboundarywaters.orst.edu provides access to the Transboundary Freshwater Dispute Database. On settlement of water disputes, see International Bureau of the Permanent Court of Arbitration (2003).

[131] UN Charter, Article 2.3: 'All Members shall settle their international disputes by peaceful means in such a manner that international peace and security, and justice, are not endangered.' Chapter IV of the UN Charter further deals with peaceful settlement of disputes.

[132] *Southern Bluefin Tuna* cases (provisional measures), New Zealand and Australia *vs.* Japan, ITLOS Nos. 3 and 4 (1999). The ITLOS allowed for provisional measures to

drological connection between salt and fresh water, for example principles governing migratory fish that spent time in both waters also present a link between international water law and the law of the sea.

Article 22 of the ECE Convention deals with the settlement of disputes. Parties 'shall seek a solution by negotiation or other means acceptable to the parties to the dispute.' A party can accept submission to the International Court of Justice or arbitration on the basis of reciprocity in accordance with the procedure set out in annex VI.[133] The ECE Convention leaves much freedom to seek a solution and accept the compulsory settlement of a dispute. The Watercourses Convention sets out options for the peaceful settlement of disputes between parties in Article 33. According to Article 33, when agreement cannot be reached by negotiation, parties are to seek good offices, request mediation or conciliation, or agree to submit the dispute to arbitration or the International Court of Justice. The parties may recognise as compulsory *ipso facto* submission to the International Court of Justice, and/or arbitration. The Watercourses Convention contains an annex on arbitration consisting of 14 articles, arranging for the arbitration pursuant to Article 33, if not otherwise agreed by the Parties. In case the aforementioned is unsuccessful, Article 33 furthermore arranges that upon request the dispute shall be submitted to impartial fact-finding, unless otherwise agreed. The report of the fact-finding commission, submitted to the parties, is to be considered in good faith by the parties. On the one hand, the peaceful settlement of disputes is well provided for in and encouraged by both conventions. On the other hand, in the final analysis, neither the ECE Convention nor the Watercourses Convention can compel states to settle their disputes.

Cooperation in managing freshwater resources reflects the fact that sustainable water management is a global issue involving all parties. A limitation to present instruments of cooperation is that the joint bodies established to facilitate cooperation vary considerably in scope and instruments, which does not necessarily ensure an integrated approach. Furthermore, better coordination between various bodies is needed. Many joint bodies might become more effective if their mandates were strengthened and an international body set up to coordinate an integrated approach.

---

protect the tuna stock because of scientific uncertainty on the conservation of tuna stocks. Case concerning *Land Reclamation by Singapore in and around the Straits of Johor* (provisional measures), Malaysia *vs.* Singapore, ITLOS No. 12 (2003). See www.itlos.org.

[133] Declarations to the ECE convention of Austria (made upon ratification, 25 July 1996) and Liechtenstein (made upon accession, 19 November1997): Both means of dispute settlement (article 22, para. 2) are accepted as compulsory on the basis of reciprocity. Declaration Netherlands to the ECE convention (made upon signature, 18 March 1992, confirmed upon acceptance, 14 March 1995): Where a dispute is not resolved in accordance with article 22(1), submission to the ICJ and Arbitration (in accordance with Annex IV) are accepted on the basis of reciprocity.

Water problems are embedded in all sectors of society and states alone cannot perform the tasks necessary to resolve them. Principles of cooperation and dispute settlement bodies in international law are largely addressed to states and intergovernmental organizations; but cooperation is needed between states and non-state actors and among non-state actors as well. In cases of conflicts, non-state entities may have the possibility to take private recourse to courts and administrative tribunals.[134] Where parties to a conflict agree to submit the dispute to, for example, the Permanent Court of Arbitration (PCA) or the World Bank Inspection Panel, non-state actors can also participate or even initiate a process.[135]

3.4.3   From state interests toward common interests

For a long time, only states were acknowledged as the subjects of international law.[136] During recent decades, however, international organizations, composed of states, have become recognised as subjects of international law as well. Although the focus remains on states and intergovernmental organizations, non-state entities are entering the international law arena in various ways.[137] Individuals and peoples have been granted human rights under international law. Moreover, non-state actors are increasingly referred to in other instruments of international law. Non-state entities such as NGOs are allowed as observers, for example, to various UN organs.

The increasing importance of non-state actors is reflected in a move away from state interests toward common interests.[138] According to Hey, the common-interest normative pattern 'seeks to regulate the interests that are common to these actors or the interests of the international community.'[139] In the classical understanding of state consent, a state in principle has to consent explicitly or tacitly in order to be bound by a rule of international law. In the case of consent to community interests, the outcome is unknown at the time of consent.

Where international law does not adequately anticipate the rise of community interests, a clash between state interests and community inter-

---

[134] McCaffrey (2001), p. 444.

[135] Of special interest to water are the Optional Rules for Arbitration of Disputes Relating to Natural Resources or the Environment of the PCA, see www.pac-pca.org. The World Bank Inspection Panel was established by identical resolutions IBRD Res. 93-10 and IDA Res. 93-6 adopted on 22 September 1993, see www.inspectionpanel.org.

[136] Shaw (1997), Chapter 5.

[137] On the implications for the concept of sovereignty of states, Schrijver (2000), argues that although the concept is evolving, the principle remains a corner stone of international law. On non-state actors, see Gupta (2003).

[138] When speaking of state interests, it is to be noted that a government often represents plural interests of various departments and sectors of society and cannot always be seen as one entity.

[139] Hey (2003), p. 11. See Hey (2003) in general for an analysis of community interests.

ests could occur.[140] According to Hey, 'the inter-state nature of the current international legal system entails that that system is ill-equipped to translate social relationships that are arising as a result of globalization into legal relationships.'[141]

Common interests transcend traditional international law and are reflected in the concept of community of interests. Although community interests may not be the same as community of interests, the concepts are closely related in the sense that they both represent a larger, more collective interest. As stated by Birnie and Boyle, common management 'represents a community of interest approach which goes beyond the allocation of equitable rights, however, and opens up the possibility of integrated development and international regulation of the watercourse environment.'[142] The community of interests reinforces the trend under discussion.[143]

According to McCaffrey, the community of interests even has its antecedents in Roman law. Further, it 'derives from the idea that a community of interests in the water is created by the natural, physical unity of a watercourse.'[144] Community of interests adds to the principle of limited territorial sovereignty and integrity by emphasising responsibilities toward, e.g., the high seas and non-riparian states and by calling for joint action.[145]

Although no explicit references to community of interests in modern treaties can be found, the many references to shared resources and the establishment of joint bodies can be argued to approach the principle.[146] At the regional level, community of interests can be found in the Protocol on Shared Watercourse Systems in the Southern African Development Community (SADC Protocol), obliging the member states in Article 2.2 to respect and abide by the principles of community of interests in the equitable utilization of shared watercourse resources.[147]

---

[140] Hey (2003), p. 5, states that efforts to address community interests introduce systematic change into the existing international legal system that meets with opposition but does not prevent normative development, resulting in situations of normative pluralism.

[141] Hey (2003), p. 5.

[142] Birnie and Boyle (2002), p. 304. On common management see Birnie and Boyle (2002), pp. 304-305.

[143] McCaffrey (2001), pp. 149-171.

[144] McCaffrey (2001), pp. 149-150.

[145] Cf. McCaffrey (2001), pp. 168-169. See Section 5.2.2 of this book on the principle of limited territorial sovereignty and integrity.

[146] See McCaffrey (2001), p. 156 and pp. 158-160.

[147] Protocol on Shared Watercourse Systems in the Southern African Development Community Region, 28 August 1995, entered into force 29 September 1998. The Revised Protocol of 7 August 2000, when entering into force, will replace the Protocol and does not have a provision similar to Article 2(2).

In relation to navigation, the principle of community of interests was confirmed between riparian states in the 1929 *River Oder* case, in which the PCIJ stated, for example:[148]

> This community of interest in a navigable river becomes the basis of a common legal right, the essential features of which are the perfect equality of all riparian States in the use of the whole course of the river and the exclusion of any preferential privilege of any one riparian State in relation to the others.

Following the above quotation from the *River Oder* case, the ICJ in the *Gabcíkovo-Nagymaros* case continues:[149]

> Modern development of international law has strengthened this principle for non-navigational uses of international watercourses as well, as evidenced by the adoption of the Convention of 21 May 1997 on the Law of the Non-navigational Uses of International Watercourses by the United Nations General Assembly.

It therefore refers to community of interests as the basis for equitable and reasonable utilization.[150]

Both community interests and community of interests express the call for an alternative to (solely) state interests. The transboundary nature of water, globalisation and increasing water problems are likely to enforce this call. International water law has to respond to it if sustainable development of freshwater resources is the aim.

3.5   Conclusions

The concept of sustainable development has gained much support as reflected in an international law on sustainable development. Sustainable development is widely accepted as an overarching objective of the international community. The international law on sustainable development is emerging and builds on human rights law, international development law and international environmental law. All three generations of human rights include provisions important to water use, emphasising the indivisible nature of human rights. Water is part of the environment and the relationship between water and development appears also fully acknowledged. However, the recent developments in international environmental law are not matched by international development law, which appears to have come

---

[148] PCIJ, 1929, in its judgement on the territorial competence of the River Oder Commission, PCIJ Series A No. 23, pp. 27-28.

[149] *Gabcíkovo-Nagymaros* case, para. 85.

[150] See Section 3.3.2 on equitable and reasonable utilization.

largely to a standstill. The objective of sustainable development and the principles of the emerging international law on sustainable development need to be further taken into account by international water law.

The water crisis calls for an integrated approach, also within international law. Modern international water law has taken an important step forward in adopting an integrated approach and taking into account all elements of sustainable development. The identified key principles of international water law – equitable and reasonable utilization, the obligation not to cause significant harm and the duty to cooperate – are capable of accommodating social, economic and ecological interests. However, international water law remains to a certain extent fragmented, ambiguous in its outcomes and focused on state interests.

The principle of equitable and reasonable utilization and its outcomes are to be adjusted to the goal of sustainable development. First, sustainable development should be set as its goal. Second, further protection of vital human needs, ecosystem protection and sustainability is needed. Third, cooperation is to be enhanced. And fourth, participation in the process is to be extended beyond (riparian) states.

The trends toward integration, cooperation and community interests can contribute to sustainable development and are to be strengthened in international law and facilitated by joint bodies. International law can facilitate cooperation between parties if it safeguards the various social, economic and ecological interests in principles and can provide for a framework for water management that contributes to a balanced and integrated approach to the combination of principles. The social, economic and ecological interests appear to respectively focus on access to water, control over water and protection of water. These three key issues are separately analysed in the following Part II.

PART II. WATER WITHIN THE PILLARS OF SUSTAINABLE DEVELOPMENT

4. Water as a social good

4.1 Access to water

This Chapter addresses water as a social good and focuses on access to water. As formulated in the report of the thematic sessions on valuing water at the Second World Water Forum, the challenge is:[1]

> To manage water in a way that reflects its economic, social, environmental and cultural values for all its uses, and move towards pricing water services to reflect the cost of their provision. This approach should take account of the need for equity and the basic needs of the poor and vulnerable.

Access to water is a condition for meeting the basic needs of all people now and in the future. The importance of access to water for sustainable development is reflected in the key commitment made at the WSSD to halve, by the year 2015, both the number of people without access to safe drinking water and the number of people who do not have access to basic sanitation.[2] Access to water refers to both physical and economic accessibility to adequate quantities and qualities of water resources required for people to meet their basic needs.[3] Affordability of water does not refer to free supply but implies that the extent of cost-recovery will depend on the ability to pay.

This brings us to the status of access to water in international water law. The ECE Convention and the Watercourses Convention emphasise the importance of meeting vital human needs.[4] According to Article 10 of

---

[1] See Report of the Thematic Sessions on Valuing Water, World Water Council (2000), pp. 55-56. In this session water as a basic human right was recognised but linked to the acknowledgement that it should not be provided free of charge.

[2] The commitment to halve the proportion of people without access to safe drinking water is a reaffirmation of the Millennium Development Goal. Announcements made during the WSSD on water and sanitation include that of the United States to invest 970 million dollars on water and sanitation projects over the next three years, the "Water for Life" initiative of the European Union that seeks to engage partners to meet goals for water and sanitation, primarily in Africa and Central Asia, a grant of 5 million dollar provided by the Asian Development Bank to UN Habitat and another 500 million dollar in fast-track credit for the Water for Asian Cities Programme, and 21 other water and sanitation initiatives with at least 20 million dollar in extra resources received by the UN. Other activities are to be expected, *e.g.*, related to the International Year of Freshwater 2003.

[3] 2002 General comment No. 15, para. 12(c), p. 6, see note 16.

[4] Article 3.20 Berlin Rules defines vital human needs as 'waters used for immediate human survival, including drinking, cooking, and sanitary needs, as well as water needed for the immediate sustenance of a household.'

the Watercourses Convention, no use is granted inherent priority over any other use. It also states, however, that in the case of conflict vital human needs call for special attention. In Section 5.2.3 of this study, the priority of uses of water is further elaborated, including water use for basic needs. Special regard for basic human needs can also be found in the ECE Protocol on Water and Health to the ECE Convention which aims to promote the protection of human health and well-being, including the goal of access to drinking water and the provision of sanitation for everyone within ECE countries.[5] Access to water is thus promoted by the Watercourses and ECE Conventions but is referred to in terms of basic need instead of a right and is not granted inherent priority. Nevertheless, at the national and community level there exists a long practice of granting people access to water for their basic needs in various cultures and regions.[6]

A right of access to water is much more strongly expressed in international soft law instruments and other documents. The 1977 Mar del Plata Action Plan states that 'all peoples, whatever their stage of development and their social and economic conditions, have the right to have access to drinking water in quantities and of a quality equal to their basic needs.'[7] According to the 1992 Dublin Statement water should be recognized as an economic good, but with recognition of the basic right of all human beings to have access to clean water and sanitation at an affordable price.[8] Chapter 18 of Agenda 21 emphasises the priority to be given to the satisfaction of basic needs and to the safeguarding of ecosystems, without being very specific. In the Johannesburg Declaration reference is made to water as a basic requirement.[9]

---

[5] Under Article 4 of the ECE Protocol on Water and Health, parties are to take measures such as to prevent, control and reduce water-related disease. In Article 4 of the Protocol Parties are to take measures in order to ensure adequate supplies of wholesome drinking water and adequate sanitation.

[6] For example, in traditional Islamic law.

[7] *Report of the United Nations Water Conference, Mar del Plata*, 14-25 March 1977, United Nations Publications: New York, E/77/II/A/12. See on the right of access to water also the work of the Second International Water Tribunal, based on the Declaration of Amsterdam which was prepared by an international group of lawyers and states: 'All members of present and future generations have the fundamental right to a sustainable livelihood including the availability of water of sufficient quality and quantity.' The Tribunal consisted of an international jury which heard 22 cases and whose pronouncements have no legally binding status, but resulted in four case books on dams, pollution, mining and management. See on the Tribunal also Hey and Nollkaemper (1992).

[8] In the Ministerial Declaration of The Hague the importance of access to sufficient safe water at an affordable cost and of managing water in a way that reflects its economic, social, environmental and cultural values for all its uses is acknowledged. See *Ministerial Declaration of The Hague on Water Security in the 21st Century*, agreed to on 22 March 2000, pp. 1-2, see http://www.worldwaterforum.net/ Ministerial/declaration.html.

[9] Johannesburg Declaration, para. 18: 'to speedily increase access to basic requirements such as clean water'.

Although the commitments made, such as reflected in Agenda 21 and the Johannesburg Declaration, emphasise the importance attached to access to water by states and contribute to its emergence as a right under international law, it is difficult to argue that this alone indicates an *opinio iuris* of states in favour of a right of people to water. According to the commentary to Article 17 of the Berlin Rules on the right of access to water, there is increasingly support in legal instruments for such a right.[10]

The sections below examine instruments that could promote access to water for all people and will present the following reasoning. The right of access to a certain quantity and quality of water for all would pose a strong legal obligation on states if considered a human right. The actual realisation of access to water would require the eradication of poverty, for example, in order to empower people to protect their basic needs and claim their rights. Furthermore, for access to water to be consistent with the aim of sustainable development (in which basic human needs and the environment are protected) it has to address intra- and intergenerational equity as well as the basic water needs of fauna and flora.

4.2 A human right to water

This section first discusses whether or not a separate right to water exists within the body of human rights law.[11] The right to water would provide a means for people to claim access to water for their basic needs. The recognition of access to fresh water as a civil and political right of individuals would oblige states to provide their citizens with, or at least not to obstruct, such access.[12] A right of access to fresh water as an economic, social and cultural right for individuals would provide a strong legal ground for addressing governments as well, entailing a duty of states to gradually implement the right depending on the means available to a state. The third category, collective rights, could provide peoples with a right to water to be respected by their state.

A right to water is not explicitly acknowledged as a human right in the Universal Declaration or the 1966 Covenants. As with a right of access to water, the existence of a human right to water under customary international law may be developing but for now remains debatable. However, human rights treaties and especially the right to an adequate standard of

---

[10] ILA Committee on Water Resources Law (2004), p. 23.

[11] On a human right to water, see Salman and McInerney-Lankford (2004), Scanlon, Cassar and Nemes (2004), Hildering (2004), WHO (2003), Gleick (2000), Smets (2000), Chapter 1, and McCaffrey (1992).

[12] However, Hunter, Salzman and Zaelke (2002), p. 826, note that: 'Such obligations are hard to enforce, but serve to highlight the importance of water to the poor and establish the use of water for direct human consumption and for food production as highest priority uses.' The dilemmas accompanying a human right of access to fresh water, such as reconciling it with its ecological role, are discussed as well.

living, can be argued to imply a human right to water and international support for this argument appears to be increasing, including organizations such as the Green Cross and the WHO.[13] Remarkably, the South African constitution includes a human right to water.[14]

Within the larger body of human rights law, reference is made to a right of access to water in various respects. The special position of drinking water is also underlined within humanitarian law, which, for example, provides that supplies necessary for survival, such as drinking water facilities, in principle are not to be attacked in the case of international armed conflict.[15]

Of great importance is the recent acknowledgement of a human right to water by the Committee on Economic, Social and Cultural Rights in General Comment No. 15 of November 2002.[16] When parties to the ICESCR report to this Committee, they thus have to deal with water as a human right under the treaty and therefore progressively ensure access to water to their population without discrimination. Although not legally binding on its own merits, the Comment is undoubtly an authoratitive statement and assessment of the status of international law.[17] The Comment not only identifies a right to water, it furthermore elaborates upon its content and the obligations it poses on states in relation to protection and fulfilment of the right to water, provides illustrations of violations of the right to water, and discusses implementation at the national level and the

---

[13] In 2004, organizations including the Green Cross International have prepared a document on principles of a convention on a right to water, explicitly stating that it is a human right, see www.greencrossinternational.net/Tools/petition/principes.html. See also WHO (2003). Gleick (2000), p. 1-17, argues that access to a basic water requirement is a fundamental human right implicitly supported by international law, declarations and state practice and further stating that a transition is underway making a right to water explicit. According to Hunter *et al.* (2002), p. 826: 'Access to safe and affordable water for basic needs is increasingly being viewed as a human right.' In ACUNS, *Plan of Action for Johannesburg: The Development-Environment Nexus*, distributed at PrepCom III, UN Headquarters, 28 March 2002, the human right to drinking water and the human right to water for all peoples are mentioned. At both the 2nd and 3rd World Water Forums, access to water was considered by many to constitute a human right. The parallel Ministerial Conferences, however, did not include a human right to water in their Declaration.

[14] See Section 8.4 of this book.

[15] Article 54(2) of the Protocols Additional to the Geneva Conventions of 12 August 1949, and relating to the Protection of Victims of International Armed Conflicts (Protocol I), 8 June 1977, prohibits the attack, destruction, removal or rendering useless of objects indispensable to the survival of the civilian population such as drinking water installations and supplies and irrigation works, see also Grünfeld (1994), p.76.

[16] General comment No. 15 (2002), The Right to Water, Substantive Issues Arising in the Implementation of the International Covenant on Economic, Social and Cultural Rights, UN Economic and Social Council, *UN Doc.* E/C.12/2002/11 (26 November 2002), see www.unhchr.ch/html/menu2/6/gc15.doc.

[17] See ILA Committee on International Human Rights Law and Practice (2004) for an elaborate analysis of the legal status of UN Committee documents.

obligations of non-state actors. The first paragraph of the Comment states that:

> Water is a limited natural resource and a public good fundamental for life and health. The human right to water is indispensable for leading a life in human dignity. It is a prerequisite for the realization of other human rights.

According to the Comment (par. 2), everyone is entitled 'to sufficient, safe, acceptable physically accessible and affordable water for personal and domestic uses.' The interrelationship between a right to water and other human rights is identified, whereby the right to an adequate standard of living (Article 11 ICESCR) and the right to the highest attainable standard of health (Article 12 ICESCR) are specifically emphasised. In the allocation of water, the Comment states that priority is to be given to the right to water for personal and domestic uses, as well as water resources required to prevent starvation and disease and water resources required to meet the core obligations of each of the Covenant rights (para. 6). Another remarkable statement in the Comment can be found in paragraph 11: 'Water should be treated as a social and cultural good, and not primarily as an economic good.' The paragraph continues to state that: 'The manner of the realization of the right to water must also be sustainable ensuring that the right can be realized for present and future generations.' The need for international cooperation between states as well as non-state actors is also emphasised.

In its work, the Sub-Commission for the Promotion and Protection of Human Rights has also dealt with a human right to water. Of special interest is the working paper of the Special Rapporteur for the Sub-Commission, Mr. El Hadji Guissé, on the right of access of everyone to drinking water supply and sanitation services (further elaborated in General Comment No. 15), which states: 'Since drinking water is a vital resource for humanity, it is also one of the basic human rights.'[18] Another document referring to water as a human right is the Earth Charter, which includes a human right to drinking water, clean air, food security, shelter and safe sanitation (principle 9).[19]

The status of access to water as a separate human right remains debatable, mainly because states have not clearly expressed their position on the subject, whether by means of treaty or by a UN General Assembly declara-

---

[18] Working paper of the Commission on Human Rights' Sub-Commission on Prevention of Discrimination and Protection of Minorities (now called Sub-Commission for the Promotion and Protection of Human Rights), E/CN.4/Sub.2/1998/7, 10 June 1998, para. 3.

[19] The Earth Charter is a private initiative launched on 29 June 2000, dealing with respect and care for the community of life, ecological integrity, social and economic justice, and democracy, non-violence and peace. See ILA Committee on Legal Aspects of Sustainable Development (2000), pp. 12-13.

tion. Their caution in accepting a potentially new human right can be understood, since proliferation in this field holds the danger of undermining the authority of the existing human rights. Nevertheless, the option of new human rights should not to be excluded. As stated by Eleanor Roosevelt: 'We will have to bear in mind that we are writing a Bill of Rights for the world, and that one of the most important rights is opportunity for development. As they grasp that opportunity, they can also demand new rights, if these are broadly defined.'[20] The crucial function of water for human life and dignity makes a right to water a more than suitable candidate to be agreed to as a human right.[21] The explicit acknowledgment by the international community of a universal and separate human right to water would increase clarity regarding its status within international law and underline its importance for human dignity.

Apart from the question of a separate human right or not, access to water imposes a condition for many other human rights. For example, the right of self-determination, as expressed in Article 1 of both 1966 Covenants, grants peoples the right to dispose freely of their natural resources. In the same Article it is stated that: 'In no case may a people be deprived of its own means of subsistence.' Access to water is an even more direct condition for the right to life and the right to health and the right to an adequate standard of living.[22] These will be discussed immediately below. For people to claim their right to water, they need to be able to participate in the process of decision-making over water and its management, which will be looked into at the end of Section 4.2.

4.2.1   The right to a healthy life

Article 6 of the ICCPR states that every human being has the inherent right to life. It goes on to state that this right must be protected by law and no one is to be arbitrarily deprived of it. The right to life constitutes *jus cogens*. The right to life is not to be derogated, not even in time of public emergency (Article 4 ICCPR). According to the Human Rights Committee: 'The expression "inherent right to life" cannot properly be understood in a restrictive manner, and the protection of this right requires that States adopt positive measures.'[23]

---

[20] Cited in Nayak (1992), p. 145, referring to F. Roosevelt, 'My Day, Feb. 6, 1947' cited in M.G. Johnson, 'The contribution of Eleanor and Franklin Roosevelt to the development of international protection for human rights', *HRQ* (1987), p. 36 n. 50.

[21] See also McCaffrey (1992), p. 24, arguing that rights to sustenance concern vital human needs and are more fundamental than certain other rights under the 1966 Covenants.

[22] Gleick (2000), p. 8: 'At a minimum, therefore, the explicit right to life and the broader rights to health and well-being include the right to sufficient water, of appropriate quality, to sustain life.'

[23] Human Rights Committee (1982), General Comment No. 6 on Art. 6 ICCPR, adopted at the Sixteenth session.

According to Article 12 of the ICESCR, everyone has the right to the enjoyment of the highest attainable standard of physical and mental health.[24] Article 24 of the CRC, concerning the right of the child to the enjoyment of the highest attainable standard of health, implies that a healthy life requires a minimum access to water. Under this Article, states are obliged to take appropriate measures to combat disease and malnutrition, which include the provision of clean drinking water. In the African Charter on the Rights and Welfare of the Child, a comparable provision is included.[25]

Within the soft law category, Principle 1 of the Rio Declaration states that: 'Human beings are at the centre of concerns for sustainable development. They are entitled to a healthy and productive life in harmony with nature.' Absence of access to drinking water and sanitation directly impairs the rights to life and to health. As stated in Chapter 2 of this study, in developing countries most diseases are water related. Lack of access to clean drinking water and adequate sanitation cause thousands of people to die on a daily basis. The direct link between a healthy life and access to water is expressed in the Plan of Implementation of the WSSD which calls for: 'Increase of access to sanitation to improve human health and reduce infant and child mortality, prioritizing water and sanitation in national sustainable development strategies and poverty reduction strategies where they exist.'[26]

At the national level, the Indian jurisprudence is most notable for extending human rights to include water issues. Indian courts have acknowledged that the right to life includes a right of access to water, expanding the right to life as expressed in Article 21 of the Fundamental Rights chapter of the Constitution to include the right to potable water.[27]

It should be clear that adequate access to water poses a condition to the right to life and to health, and that without a right to water those rights would be deprived of substance.

4.2.2 The right to an adequate standard of living

The right to an adequate living standard is expressed in Article 25 of the Universal Declaration of Human Rights and in Article 11 of the ICESCR. According to Article 11.1 of the ICESCR: 'The States Parties to the present Covenant recognize the right of everyone to an adequate standard of living

---

[24] Toebes (1999), p. 255, elaborates on clean drinking water and adequate sanitation as underlying preconditions for health. At p. 270, she states that: 'Access to clean drinking water and adequate sanitation, adequate nutritious foods, prevention of occupational diseases, and a healthy environment can be considered as elements of the scope of the right to health.'

[25] Article 14 of the African Charter on the Rights and Welfare of the Child, *OAU Doc.* CAB/LEG/24.9/49 (1990), entry into force: 29 November 1999.

[26] WSSD Plan of Implementation, para. 6(m).

[27] *Attakoya Thangal* v. *Union of India* 1990 (1) KLT 580. See also Anderson (1996), pp. 214-215.

for himself and his family, including adequate food, clothing and housing, and to the continuous improvement of living conditions.' Such an adequate standard of living cannot be achieved without a minimal access to water. Moreover, although in practice water and food are often dealt with separately, it might be argued that "food" implies water. Water appears to be included in the general definition of food: "any nutritious substance that people or animals eat or drink or that plants absorb in order to maintain life and growth".[28] According to McCaffrey: 'the right to food should be interpreted as the right to receive life-sustaining nourishment, or sustenance, so that it would include the right to potable drinking water sufficient to sustain life.'[29] The right to adequate housing furthermore implies the availability of housing with facilities such as sanitation and drinking water.[30]

Other articles within human rights treaties on the standard of living include Article 14(2)h of the 1979 Convention on the Elimination of All Forms of Discrimination Against Women – referring to the right of rural women to enjoy adequate living conditions including in relation to sanitation and water supply[31] – and Article 27(3) CRC – obligating state parties in case of need to provide assistance with regard to nutrition. Moreover, the UN Charter in its preamble expresses the aim of promoting social progress and better standards of life in freedom. Article 55 of the Charter furthermore states that the UN shall promote higher standards of living and conditions of economic and social progress and development.

Like the right to life and the right to health, the right to an adequate standard of living is firmly embedded in international law and sets another obligation for states to provide people with access to sufficient water and adequate sanitation.

4.2.3   Participation

Public participation is acknowledged to be an essential part of sustainable development. According to the ILA, participation in the processes whereby decisions are made by persons likely to be affected is now a well established human right.[32] The importance of participation as a procedural right is well-formulated in Principle 10 of the Rio Declaration:

---

[28] Pearsall (1999), p. 551.

[29] McCaffrey (1992), p. 24.

[30] Westendorp (1994), p.105, referring to the General Comment 4 on the right to adequate housing of the Committee on Economic, Social and Cultural Rights, E/C.12/1991/1, mentions drinking water as one of the facilities required for a house to fit the qualification 'adequate'.

[31] Convention on the Elimination of All Forms of Discrimination against Women, New York, 18 December 1979, entry into force: 3 September 1981, 1249 *UNTS*, 13. Status as of 4 October 2004: 178 parties and 98 signatories.

[32] Commentary to Article 4 of the Berlin Rules, ILA Committee on Water Resources Law (2004), p. 12.

Environmental issues are best handled with the participation of all concerned citizens, at the relevant level. At the national level, each individual shall have appropriate access to information concerning the environment that is held by public authorities, including information on hazardous materials and activities in their communities, and the opportunity to participate in decision-making processes. States shall facilitate and encourage public awareness and participation by making information widely available. Effective access to judicial and administrative proceedings, including redress and remedy, shall be provided.

To implement sustainable development and, more importantly, to enforce a human right to water, people are to be allowed to participate in the process of decision-making and to be involved in the management of water resources. In this way, they are provided with an instrument to enforce their right to water. Involvement raises the necessary awareness of water problems and can therefore contribute to their resolution. Participation therefore not only constitutes a right but also a duty to assume responsibility. Moreover, participation requires the protection of such human rights as the freedom of expression. The principle of participation is part of treaty law and also of the emerging international law on sustainable development.

Apart from more general articles on cooperation, Article 16.1 of the ECE Convention on Watercourses obliges riparian parties to 'ensure that information on the conditions of transboundary waters, measures taken or planned to be taken to prevent, control and reduce transboundary impact, and the effectiveness of those measures, is made available to the public.'[33] The 1998 ECE Convention on Access to Information, Public Participation in Decision-making and Access to Justice in Environmental Matters (Aarhus Convention) further underlines the importance of participation.[34] The Aarhus Convention is viewed as the most important and comprehensive elaboration of Rio Principle 10.[35] Another example of a regional agreement extensively regulating participation is the EC Water Framework Directive (EUWFD).[36] The Watercourses Convention does not contain provisions on

---

[33] Article 16.1 ECE Convention furthermore states that the information to be made available to the public includes water-quality objectives, conditions required by permits issued and certain results of water and effluent sampling. Article 16.2 provides that the information must be available for inspection at all reasonable times and that copies of such information shall be obtainable by members of the public.

[34] ECE Convention on Access to Information, Public Participation in Decision-making and Access to Justice in Environmental Matters, Aarhus, 25 June 1998, entry into force: 30 October 2001, 38 *ILM* (1999) 517. Status as of 4 October 2004: 30 parties and 40 signatories. See www.unece.org/leginstr/cover.htm.

[35] Birnie and Boyle (2002), pp. 262-263.

[36] Directive 2000/60/EC of the European Parliament and of the Council of 23 October 2000 establishing a framework for Community action in the field of water policy, entry

public participation. However, in its preamble the UN Watercourses Convention addresses the valuable contribution of international organizations. In case of emergency situations, competent international organizations are to be notified and, where appropriate, to be cooperated with (Article 28). Relating to persons, natural or juridical, the Convention contains a non-discrimination clause (Article 32). Other provisions in conventions relevant to participation include Article 6 of the UNFCCC – on education, training and public awareness – and Article 19 of the UNCCD – on capacity building, education and public awareness.[37]

Turning to soft law, the documents resulting from UNCED often address both states and people. Besides Principle 10 of the Rio Declaration, participation is an important aspect throughout Agenda 21, including Chapter 18 on freshwater resources. Implementation of Agenda 21 remains primarily the responsibility of governments, but is supported by international cooperation and contributions from international, regional and subregional organizations. The Rio Declaration on Environment and Development deals with participation in many of its principles.

The ILA New Delhi Declaration contains a separate principle on participation: Principle 5 on the principle of public participation and access to information and justice.[38] According to Principle 5.1:

> Public participation is essential to sustainable development and good governance in that it is a condition for responsive, transparent and accountable governments as well a condition for the active engagement of equally responsive, transparent and accountable civil society organizations, including industrial concerns and trade unions.

Principle 5 also includes the requirements of freedom of expression and a right of access to information (Principle 5.2) and access to effective judicial or administrative procedures (Principle 5.3).

Public participation can take many forms. At present, the need for public participation in the early stages of decision-making and water management is increasingly acknowledged. This necessity follows, *inter alia*, from the need to create support in order to implement regulations. Moreover, the skills and knowledge of their environment that people at the community level hold are valuable in achieving sustainable management of freshwater resources. Participation involves reciprocity: decision-makers must not merely inform people but be informed by them as well. Public participation furthermore provides an instrument of additional supervision over the implementation of sustainable water management. The participation of the private sector can also provide additional means for implementation.

---

into force: 22 December 2000, addressed to the Member States. See on the EUWFD also Section 8.4 of this study.

[37] On the UNFCCC and participation see Gupta (2003).

[38] See www.un.org/ga/57/document.htm for the ILA New Delhi Declaration.

The participation of non-state and non-elected entities can, however, raise questions about the legitimacy of processes if, for example, selection criteria or accountability of influential participants are missing or unclear.[39] Moreover, the international community should not too easily assume that the requirements of participation are met. For example, the present conditions of the ownership of land and water often mean that groups such as women and the poor in a society are inadequately represented. Furthermore, participants may be selected by governments and/or not be a member of the group they are supposed to represent. Especially considering vulnerable groups of people, participation on paper may need to be checked to ensure that it corresponds with actual participation of that group. An additional complication is the fact that active participation costs considerable time. Some groups of people may not be able to afford this.

International law must broaden its scope and increase and formalise the participation of various non-state entities at various policy-levels if to reflect the importance of these actors. Their increasing role might need to be countered by further regulation of non-state actor responsibilities.[40] Moreover, participation of people on a more equal basis calls for measures such as to promote the eradication of poverty.

4.3   Eradication of poverty

One of the Millennium Development Goals is to halve by 2015 the proportion of people living in extreme poverty. Eradication of poverty has become one of the priorities of development, furthermore reflected in the 1995 Copenhagen Programme of Action resulting from the UN World Summit for Social Development and the 1992 UNGA Resolution declaring 17 October as International Day for the Eradication of Poverty.[41] Moreover, the period 1997-2006 is the first UN Decade for the Eradication of Poverty.[42] Poverty is often defined by income: 1 US$ per day or less.[43] The United Nations Development Programme (UNDP) addresses poverty as a denial of human rights such as the right to health and an adequate standard

---

[39] See Gupta (2003). According to Brunnée and Toope (1997), equality, transparency, justice and fairness are fundamental legal values and therefore linked to legitimacy.

[40] Many existing multilateral instruments on corporate responsibility are non-binding under international law – such as the 1976 OECD Guidelines for Multinational Enterprises (revised several times) and the voluntary UN Global Compact with transnational corporations which includes social responsibility guidelines – while the binding appropriate ILO Conventions are difficult to implement.

[41] See www.un.org/esa/socdev/wssd on the World Summit. UNGA Resolution on the observance of an international day for the eradication of poverty, *UN Doc.* A/RES/47/196 (1992). See also www.undp.org/mainundp/propoor.

[42] www.un.org/events/poverty2000.

[43] www.worldbank.org.

of living.[44] The Human Development Reports of the UNDP view poverty to include dimensions such as lack of political freedom and the inability to participate in decision-making.[45]   The reports include the Human Development Index – based on longevity, knowledge and a decent standard of living. The 1997 Human Development Report introduced the Human Poverty Index, which measures deprivation by examining illiteracy, malnutrition among children, early death, poor health care, and poor access to safe water.[46] The 1997 Report furthermore indicates policy measures for the eradication of extreme poverty, including gender equality, improvements in the management of globalisation and promotion of accountability in governance and a strong role for civil society actors. The 2003 Report elaborates upon the Millennium Development Goals.[47] Eradication of poverty is acknowledged as a goal but is not yet a well-established principle under international law.[48]

Under Principle 5 of the Rio Declaration, all states and people are to cooperate on eradication of poverty 'in order to decrease the disparities in standards of living and better meet the needs of the majority of the people of the world.' According to paragraph II.6 of the WSSD Plan of Implementation: 'Eradicating poverty is the greatest global challenge facing the world today and an indispensable requirement for sustainable development, particularly for developing countries.'

Eradication of poverty is an objective in the ILA New Delhi Declaration, formulated in Principle 2 on equity and the eradication of poverty. According to Principle 2.3, the implementation of the right to development requires the duty to co-operate in the eradication of poverty. Eradication of poverty is furthermore regarded as a minimum to be ensured by a state fulfilling its responsibility to aim for conditions of equity within its own population (Principle 2.4). Eradication of poverty is thus provided with a strong position in the Declaration but is based on the right to development, which itself has a debatable status in international law, and on a quite far-reaching interpretation of the principle of equity.[49]

While participation may require but does not safeguard the representation of all interested parties, the eradication of poverty could empower people to play their part, claim their rights and take on the responsibilities

---

[44] www.undp.org/poverty.

[45] hdr.undp.org.

[46] UNDP *Human Development Report 1997: Human Development to Eradicate Poverty*, see hdr.undp.org/reports/global/1997/en.

[47] Human Development Report 2003, *Millennium Development Goals: A compact among nations to end human poverty*, see hdr.undp.org/reports/global/2003.

[48] On poverty reduction see also the reports of the Independent Expert on Human Rights and Extreme Poverty to the UN Commission on Human Rights, *e.g. UN Doc.* E/CN.4/2003/52, see www.unhchr.ch.

[49] Section 7.2.1 of this study categorises the right to development as the principle of international law in particular combining social and economic interests. Equity is discussed in Section 4.4.

of achieving sustainable development. Capacity-building is often viewed as an essential instrument for achieving sustainable development, including the eradication of poverty. Besides the actual capabilities, it becomes clear from the foregoing paragraphs that human rights, participation and the eradication of poverty all require a state in which people can practice their rights as well as their duties. In reality, many people do not live under such circumstances. Hence, the importance of good governance for sustainable development. This brings us to the discussion of capacity-building, the involvement of certain social groups and good governance.

4.3.1 Capacity-building

Based on provisions such as Article 19.1 of the UNCCD and Article 22 of the Cartagena Protocol on Biosafety, capacity-building can be defined as the development and strengthening of relevant local and national human resources and institutional capacities through, *e.g.*, training and institution building.[50] In view of the recognition of human dignity and empowerment of people, it may be argued that capacity-building is a goal in itself. Capacity-building is part of treaty law but cannot (yet) be regarded as a general principle of international law. Elements of capacity-building however, such as the right to participate in decision-making processes and the right to education, can be regarded as well-established within international law.[51] Capacity-building is nevertheless categorised under eradication of poverty since its importance goes well beyond such rights: it is a key element to development and the overcoming of poverty. According to Agenda 21 Chapter 37.1:[52]

> Specifically, capacity-building encompasses the country's human, scientific, technological, organizational, institutional and resource capabilities. A fundamental goal of capacity-building is to enhance the ability to evaluate and address the crucial questions related to policy choices and modes of implementation among development options, based on an understanding of environmental potentials and limits and of needs as perceived by the people of the country concerned. As a result, the need to strengthen national capacities is shared by all countries.

In Chapter 18 of Agenda 21, human resources development and capacity-building are discussed extensively as part of the means of implementing

---

[50] The Cartagena Protocol on Biosafety, adopted by the Conference of the Parties to the Convention on Biological Diversity, Montreal, 29 January 2000, entry into force: 11 September 2003. Status as of 4 October 2004: 109 parties and 103 signatories. See www.biodiv.org/biosafety.

[51] On the human right to education, see Article 13 UNCESCR.

[52] Chapter 37 of Agenda 21 on national mechanisms and international cooperation for capacity-building in developing countries, see www.un.org/esa/sustdev/ documents/agenda21/english/agenda21chapter37.htm.

the integrated development and management of water resources.[53] For capacity-building to significantly contribute to the eradication of poverty, international investments are needed. Moreover, in order to enable developing countries to allocate and manage their fresh water in a sustainable fashion and to create access to a certain quantity and quality of water for people, measures such as the exchange of technology need to be given a fresh impulse.[54]

Capacity-building is specifically relevant to the empowerment of vulnerable groups of people. The WSSD Plan of Implementation in Chapter II.6 emphasises the importance of capacity-building in relation to women, indigenous people and children, including the objectives to promote equal access by women to decision-making at all levels and to land, credit and education, to develop ways to improve indigenous people's access to economic activities and to ensure primary schooling and equal access to education for children.

4.3.2   Social groups

Women, youth, indigenous people and deprived people constitute social groups in a society which are often in a particularly vulnerable position and are often part of a group facing marginalisation. Special attention to the empowerment and protection of these groups at all levels is thus called for, not only in developing countries but in developed countries as well.

Apart from their involvement as stakeholders, the knowledge of the social groups under discussion is increasingly regarded to be very valuable.[55] For example, in many regions women are the ones who deal most with water, while indigenous people often have a rich knowledge of their surrounding environment.[56] Appreciation for such knowledge is growing now that the consequences of modern industrial methods for sustainable development are becoming clear.

Apart from the ILO Convention concerning Indigenous and Tribal Peoples in Independent Countries, the need for special attention to social groups is not (yet) reflected in treaty or customary law, but is all the more part of soft law instruments such as the Rio Declaration, Agenda 21, and

---

[53] Agenda 21, para. 18.19.

[54] On transfer of technology see Banerjee (1992) who underlines the importance of access to information for such a transfer to be effective and argues in favour of the development of indigenous technological capacity.

[55] See, e.g., the 1995 Beijing Declaration resulting from the UN Fourth World Conference on Women, Beijing, September 1995, and also the UNGA Resolution on the International Decade of the World's Indigenous People, UN Doc. A/RES/48/163 (1993), referring to the holistic traditional scientific knowledge of indigenous people and their communities of their lands, natural resources and environment.

[56] On the position of women, see e.g., www.un.org/womenwatch. See Barlow and Clarke (2002), pp. 231-232, on the Indigenous peoples' Declaration of Water resulting from the 2001 Summit on Water for People and Nature.

the WSSD Plan of Implementation.[57] At the international level, the following principles of the Rio Declaration especially acknowledge such groups.

Principle 20: Women have a vital role in environmental management and development. Their full participation is therefore essential to achieve sustainable development.
Principle 21: The creativity, ideals and courage of the youth of the world should be mobilized to forge a global partnership in order to achieve sustainable development and ensure a better future for all.
Principle 22: Indigenous people and their communities and other local communities have a vital role in environmental management and development because of their knowledge and traditional practices. States should recognize and duly support their identity, culture and interests and enable their effective participation in the achievement of sustainable development.

The ILA New Delhi Declaration emphasises the 'critical relevance of the gender dimension' in its preamble and underscores the vital role of women in sustainable development and the need for their full participation in all levels of decision-making in Articles 5.1 and 6.3 respectively. The Dublin Statement stresses the central role of women in the provision, management and safeguarding of water.[58] Principle 3 of the Dublin Statement furthermore states:

This pivotal role of women as providers and users of water and guardians of the living environment has seldom been reflected in institutional arrangements for the development and management of water resources. Acceptance and implementation of this principle requires positive policies to address women's specific needs and to equip and empower women to participate at all levels in water resources programmes, including decision-making and implementation, in ways defined by them.

Although the contribution of these provisions is not to be underestimated, the actual protection of the identified social groups by international law appears to remain inadequate and, moreover, the provisions rarely address deprived people as a separate social group in need of further attention and protection. A further complication is that in practice it is often risky for vulnerable groups to stand up for themselves, for example in

---

[57] ILO Convention No. 169 concerning Indigenous and Tribal Peoples in Independent Countries, adopted on 27 June 1989 by the General Conference of the International Labour Organisation at its seventy-sixth session, entry into force: 5 September 1991. Work on the Draft United Nations Declaration on the Rights of Indigenous Peoples is continuing.
[58] On the Dublin Statement, see also Section 1.1 of this study.

situations where human rights are systematically violated. The structure of a state and the functioning of its governance are of major importance if to ensure equal opportunities of development for the social groups.

4.3.3   Good governance

Good governance can be regarded as part of sustainable development.[59] According to Article 9 of the Cotonou Agreement: 'Respect for all human rights and fundamental freedoms, including respect for fundamental social rights, democracy based on the rule of law and transparent and account-able governance are an integral part of sustainable development.'[60] Good governance is a rather fluid concept that can be defined in various ways but the contours can nevertheless be identified.   Article 9 of the Cotonou Agreement states:

> In the context of a political and institutional environment that upholds human rights, democratic principles and the rule of law, good govern-ance is the transparent and accountable management of human, natu-ral, economic and financial resources for the purposes of equitable and sustainable development. It entails clear decision-making procedures at the level of public authorities, transparent and accountable institutions, the primacy of law in the management and distribution of resources and capacity building for elaborating and implementing measures aim-ing in particular at preventing and combating corruption.

According to UNDP: 'Good governance ensures that political, social and economic priorities are based on a broad consensus in society and that the voices of the poorest and the most vulnerable are heard in decision-making over the allocation of development resources'.[61] The OECD's main ele-ments of effective systems of governance include an enterprise-based econ-omy, a competitive environment, action against corruption and organised crime, environmental management and government investment in the peo-ple. Recurring elements of good governance include transparency, a broad consensus of society, democratic accountability, the primacy of law and the prevention of and fight against corruption.[62]

---

[59] ILA Committee on Legal Aspects of Sustainable Development (2002), p. 11, and Ginther, Denters and De Waart (1995).

[60] Partnership Agreement between the Members of the African, Caribbean and Pacific Group of States of the one part, and the European Community and its Member States, of the other part, Cotonou, 23 June 2000. See www.europarl.eu.int/intcoop/acp, eu-ropa.eu.int/comm/development/body/cotonou/agreement_en, and Arts (2003).

[61] UNDP (1997), p. 3.

[62] See on corruption the UN Convention against Corruption, 21 November 2003, *UNGA Res.* 58/4 (not yet in force). The link with sustainable development is made explicit in its preamble: 'Concerned about the seriousness of problems and threats posed by corruption to the stability and security of societies, undermining the institu-

Good governance is part of treaties such as the Treaty of Cotonou and its predecessor Lomé IV-bis, posing a condition on development aid, and can be argued to constitute an emerging principle of international law. According to Principle 6 of the ILA New Delhi Declaration, the principle of good governance 'is essential to the progressive development and codification of international law relating to sustainable development.' The principle of good governance is said to commit states and international organizations to adopt transparent decision-making, take measures to combat corruption and observe the rule of law. According to Principle 6.2, civil society and NGOs are entitled to good governance by states and intergovernmental organizations and non-state actors have to be accountable. Principle 6.3 calls for Corporate Social Responsibility (CSR) and socially responsible investments (SRI) as elements of good governance.[63]

It is desirable to strengthen the position of a principle of good governance under international law as formulated by the ILA. In providing aid under the condition of good governance, usually defined by the developed countries, reciprocity would require that developed countries look into their own governance as well as the governance of developing countries. Nevertheless, institutions and the changing content of good governance have often reflected global power relations rather than necessarily promoted sustainable development. This again emphasises the importance of representative participation of states and non-state actors.[64]

4.4   Equity

Equity is a general principle of international law that plays an important part in both the international law of the sea (*e.g.* in Articles 59, 74 and 83 UNCLOS) and international water law, especially in the principle of equitable and reasonable utilization.[65] Equity can be said to be based on ideas of fairness and justice.[66] The ICJ in the *Tunisia/Libya Continental Shelf* case states that: 'equity as a legal concept is a direct emanation of the idea of justice'.[67] The ICJ furthermore states in the *Libya/Malta* case:[68]

---

tions and values of democracy, ethical values and justice and jeopardinzing sustainable development and the rule of law'. Anti-corruption is also emphasised in IMF (1997), 'Guidelines Regarding Governance Issues', *IMF Survey*, 5 August 1997 and by Transparency International, an NGO focussing on combatting corruption, see www.transparency.org.

[63] www.triple-p.org and www.irene-network.nl.

[64] Gupta (2002).

[65] See on equity, *e.g.*, Shaw (1997), pp. 82-86, and on equity in water treaties, *e.g.*, Giordano and Wolf (2001). See on the sources of international law Section 1.4 of this book and Section 3.3.2 on the principle of equitable and reasonable utilization.

[66] On fairness, see Franck (1995).

[67] *Tunisia/Libya Continental Shelf* case, ICJ Reports 1982, pp. 18, 60.

Thus the justice of which equity is an emanation, is not an abstract justice but justice according to the rule of law; which is to say that its application should display consistency and a degree of predictability; even though it looks with particularity to the peculiar circumstances of an instant case, it also looks beyond it to principles of more general application.

Other cases referring to equity include the *Diversion of Water from the Meuse* case, the *North Sea Continental Shelf* cases and the *Barcelona Traction* case.[69] Equity is often applied to adapt, complement or correct applicable international law and could, for example, be applied to assure the provision of access to water for all interest groups. Equity as a distinct principle of international law can be regarded as part of the emerging international law on sustainable development, as laid down in Principle 2 of the ILA New Delhi Declaration. According to Principle 2.1: 'The principle of equity is central to the attainment of sustainable development.' Equity in Principle 2.1 is understood as referring to intra-generational equity and intergenerational equity. If sustainable development is to be promoted, intra-generational as well as intergenerational equity have to be taken into account (*cf.* Principle 2.2). An emerging issue is equity between species. These three applications of equity will now be discussed.

4.4.1   Intra-generational equity

The significance of reducing poverty for sustainable development in developing countries has already been stressed. Inequity within the present generation has been addressed throughout previous chapters, referring, for example, to the Rio Declaration. Although not explicitly referred to as intra-generational equity, such elements can be regarded as part of this concept.[70] Principle 2.1 of the ILA New Delhi Declaration refers to intra-generational equity as 'the right of all peoples within the current generation of fair access to the current generation's entitlement to the Earth's natural resources'. Other instruments which reflect the notion that intra-generational equity is part of international law include, for example, the Climate Change Convention and the Biodiversity Convention.

Intra-generational equity requires capacity-building at the community and national level, transfer of technology among states and with other parties, and common but differentiated responsibilities at the international

---

[68] *Case Concerning the Continental Shelf* (Libyan Arab Jamahiriya/Malta), ICJ Judgement of 3 June 1985, para. 45, see www.icj-cij.org.
[69] *Diversion of Water from the Meuse* case (Holland/Belgium), 1937, PCIJ Series A/B, No. 70. *North Sea Continental Shelf* cases (West Germany/Holland/Denmark), ICJ Reports 1969. *Barcelona Traction* case, ICJ Reports 1970.
[70] Birnie and Boyle (2002), pp. 91-92.

level. In preserving our common heritage, different countries have different obligations for various reasons such as the level of pressure exerted on this heritage by countries, for example, through pollution.[71] Investment in, *e.g.*, flood prevention and the conservation of wetlands in developing countries may call for the assistance of other states. It is not only the long-term self-interests of the international community which demand that we should not exhaust our natural resources, such as water, but also more immediate interests such as the solution of pressing situations that cause human suffering within the present generation.

4.4.2 Intergenerational equity

Intergenerational equity is a key element of sustainable development. It provides the basis to take future generations of humankind into account and ensuring that their needs can still be met, as reflected in the Brundtland definition of sustainable development.[72] Principle 2.1 of the ILA New Delhi Declaration refers to intergenerational equity as 'the right of future generations to enjoy a fair level of the common patrimony'.[73]

Many freshwater problems have their effect over a long period of time and over a large area. Already preventing many of the present generation from fulfilling their needs, these problems may thus constitute an even greater threat to the needs of future generations. As formulated by Brown Weiss: 'The pollution of both surface and ground water supplies poses the most serious problems of equity between generations.'[74] The degradation of water quality, the depletion of freshwater resources and the large-scale diversion of waters are identified by Brown Weiss as posing problems for future generations.[75]

Intergenerational equity is referred to, explicitly or by reference to human responsibility for the environment, by, *e.g.*, the 1972 Stockholm Declaration, the 1992 Rio Declaration and the 1992 Convention on Climate Change.[76] At the regional level, the 1968 African Convention on the Conservation of Nature and Natural Resources is one of the earliest treaties to include such reference.

In the *Gabcíkovo-Nagymaros* case, in referring to its 1996 Advisory Opinion on the *Legality of the Threat or Use of Nuclear Weapons*, the ICJ found that: 'the environment is not an abstraction but represents the living

---

[71] See Section 7.2.3 of this study for further analysis of the principle of common but differentiated responsibilities.

[72] For an extensive elaboration of intergenerational equity, see Brown Weiss (1989). See also Birnie and Boyle (2002), pp. 89-91.

[73] See also Brown Weiss (1989), p. 289.

[74] Brown Weiss (1989), p. 232.

[75] See Brown Weiss (1989), pp. 232-247. She recommends a basin-wide ecosystem approach to these problems.

[76] 1972 Stockholm Declaration, Principles 1 and 2. Rio Declaration, Principle 3. 1992 Convention on Climate Change, Article 3(1).

space, the quality of life and the very health of human beings, including generations unborn.'[77] The ICJ therefore recognises the importance of the environment for present as well as future generations of humankind. A remarkable case at the national level is the *Minors Oposa* case in the Philippines Supreme Court.[78] In this case, an NGO was allowed to represent children and generations yet unborn and to (successfully) challenge the large-scale destruction of Philippine rain forests.

The importance of the principle of intergenerational equity to sustainable development appears uncontroversial, and it can be viewed as *de lege ferenda*, but it seems difficult to argue that it has already emerged as a general principle of international law.[79]

4.4.3   Equity between species?

The recognition of the rights of other living species is a natural corollary to the recognition of rights of future generations.[80] The need to protect animals not only for functional reasons but also for their intrinsic value is increasingly acknowledged, for example, through the 2003 International Conference on Animal Welfare in Manila.[81]

In the sense that rights of other species refer to duties of humans to protect fauna and flora, they can be identified in many international agreements, including the 1946 International Convention for the Regulation of Whaling, the 1971 Ramsar Convention, the 1973 Convention on International Trade in Endangered Species of Wild Fauna and Flora (CITES), the 1979 Convention on the Conservation of Migratory Species of Wild Animals, the 1980 Convention for the Conservation of Antarctic Marine Living Resources, the 1992 Convention on Biological Diversity, and the 1995 UN Agreement Relating to the Conservation and Management of Straddling Fish Stocks and Migratory Fish Stocks.[82]

---

[77] ICJ, *Gabcíkovo-Nagymaros* case, para. 53.

[78] Supreme Court of the Philippines, *Minors Oposa v. Secretary of the Department of Environment and Natural Resources* (Juan Antonio Oposa and others v. The Honourable Fulgencio S. Factoran and others), case of 30 July 1993, 33 *ILM* (1994), 173.

[79] Brown Weiss (1989) argues that intergenerational equity is already established within international law, as opposed to Boyle and Freestone (1999), p. 13, who acknowledge elements such as avoidance of irreversible harm and a more general responsibility of humankind, but do not adhere to the existence of generational rights.

[80] See on animal rights, *e.g.*, Birnie and Boyle (2002), pp. 556-559.

[81] The Conference was visited by 22 government delegations and was organised by, among others, the World Society for the Protection of Animals (WSPA), whose work further includes a Universal Declaration for the Welfare of Animals, see www.wspa-international.org.

[82] International Convention for the Regulation of Whaling, Washington 1946, 161 *UNTS* 72, entry into force: 10 November 1948, amended version of 1956, entry into force: 4 May 1959, see iwcoffice.org. International Convention on International Trade in Endangered Species of Wild Fauna and Flora (CITES), Washington 1973, entry into force: 1 July 1975, 993 *UNTS* 243, and 12 *ILM* (1973), 1085. See www.cites.org. Con-

Biodiversity refers to the range and the number of the various species of flora and fauna. Currently, the loss of biodiversity is taking place at a rate that could halve the number of existing species in a single human generation. In other words, the present human way of life is leading to enormous loss and extinction of animals and plants. Overemphasising the freedom of human behaviour can generate doom scenarios not only for ecosystems but also for the human species itself. It is, moreover, a legally sound principle that rights need to be accompanied by duties. The urgency of (re-)discovering a balance with our surroundings has become increasingly widely acknowledged and requires the interests of other species to be taken into account when allocating freshwater resources.[83]

4.5   Conclusions

The efforts needed to implement the commitments made relating to water as a social good and to provide basic access to water for all people, for example at Johannesburg, are manifold. First of all, a human right to water seems to be implied in the body of human rights law. It can be argued therefore that the state has a duty to respect, protect and/or fulfil access to water. The duty to respect the right to water means that a state should not deprive people of their access to water. The duty to protect the right to water implies that a state is obliged to protect people's right to water from interference from third parties. The duty to fulfil the right to water means that a state is under the duty to facilitate access to water, taking into account the available means. The explicit declaration of a human right to water by the Committee on Economic, Social and Cultural Rights is a notable step forward and strengthens the instrument of international law in securing access to water. A human right provides a pressing argument to give priority to basic access in the allocation of water. Access to fresh water in order to fulfil basic human needs is also protected by the body of human rights law as a condition for the fulfilment of other human rights. Clarity upon the status of a right to water would nevertheless still be served by further affirmation by states.

Second, international law does not yet sufficiently reflect the necessity and reality of non-state actor participation. The enforcement of a right to

---

vention on the Conservation of Migratory Species of Wild Animals, Bonn, 23 June 1979, entry into force: 1 November 1983, 19 *ILM* (1980), 15. Convention for the Conservation of Antarctic Marine Living Resources, Canberra, 20 May 1980, entry into force: 7 April 1982, 19 *ILM* (1980), 837. 1995 Agreement for the Implementation of the Provisions of the Convention relating to the Conservation and Management of Straddling Fish Stocks and Highly Migratory Fish Stocks, entry into force: 11 December 2001, 34 *ILM* (1995), 1542.

[83] See Brown Weiss (1989), p. 37: 'To derive the principles of intergenerational equity, it is necessary to return to the underlying purpose of our stewardship of the planet: to sustain the welfare and well-being of all generations.'

water would be served by strengthening public participation within international law. The importance of representative participation of all interested parties for sustainable development of freshwater resources is reaffirmed by this Chapter.

Third, the realisation of access to water calls for further eradication of poverty and application of equity. Access to water needs to be safeguarded for both present and future generations through the principle of equity. On the one hand, it can be argued that especially in developed countries it is time for other concerns such as the environment to override economic interests in the use of freshwater resources. On the other hand, many countries in the world have not yet reached that stage of development. In their case it is desirable to increase the weight given to principles of international development law, such as the right to development and the right to an adequate standard of living. In this way, the importance of the environment can be acknowledged while at the same time securing the right to development in allocating freshwater resources, in fairness to future *and* to present generations. Although the importance of flora and fauna has received increasing consideration, the intrinsic value of the environment and the interests of species other than humans are only just being acknowledged as coming within the province of international law.

Fourth, other principles of international law can contribute to the implementation of access to water and the rights and duties discussed in this Chapter.[84] Sovereignty over natural resources, the duty to cooperate, the common concern of humankind, and common but differentiated responsibilities are of specific importance in enabling states, and the world at large, to grant people access to drinking water and sanitation in line with commitments made such as by the Millennium Development Goals. Securing a basic access to water for all people requires an integrated approach that converges several chapters of international law: access to water can be protected by human rights and peoples' rights and implemented by providing people with actual control over water and qualifying state sovereignty by duties. Control over water is subject of analysis in the next Chapter.

---

[84] These principles are categorised in other chapters in line with the methodology elaborated in Section 1.4.

5.   Water as an economic good

5.1   Control over water

Having discussed water as a social good, we now turn to water as an economic good, *i.e.* in the context of the economic pillar of sustainable development.[1] This Chapter presents the arguments for and against the treatment of water as an economic good and analyses ways for international law to make economic incentives work for sustainable development.[2]

Fresh water is scarce and therefore an economic good. Scarcity of fresh water is not normally viewed as absolute but is linked to the distribution of water. Scarcity includes regional scarcity, scarcity for part of the population, scarcity of a certain quality of water, and scarcity over certain periods of time.[3] There are multiple causes of the scarcity of water, such as mismanagement, a growing population, unequal distribution causing local scarcity, pollution causing scarcity of a certain quality of water and climate change.[4] One of the ways to manage water effectively and efficiently is to deal with water as an economic good.[5] The Dublin Statement declares:[6]

---

[1] This Chapter is based on Hildering (forthcoming). See also Savenije and Van der Zaag (2002), Savenije (2002), Green (2000), Van den Bergh (1999) and (1996), and Weiss, Denters and De Waart (1998).

[2] The arguments also reflect the opposite views within economics: 'Panglossian' *vs.* pragmatic approaches. See for a thorough elaboration upon such arguments, Green (2000), stating at p. 211: 'Panglossians believe that they have the answer whereas Pragmatists believe that economics offers a way of finding the answer by first understanding what a decision involves. Panglossians emphasise economic theory; Pragmatists emphasise analysis.'

[3] According to Hirji and Grey (1998), p. 83: 'Competition for water between sectors and countries is becoming more intense even in countries with relatively abundant supplies, such as Ghana, and water pollution and environmental degradation are intensifying as economies grow.' On scarcity of water see also Brans, De Haan, Nollkaemper and Rinzema (1997).

[4] Additional causes include the concentration of people away from water resources and technology connecting many people to the water network leading to a higher per capita demand. See Dalhuisen, De Groot and Nijkamp (2000), p. 3. On population growth and water see Villiers (2001), pp. 306-309.

[5] For example, I. Serageldin, vice-president on environment issues of the World Bank, cited in De Haan (1997), p. 245, stated: 'The saving grace for future wars over water would be if the universal natural resource water were to assume its proper place as an economically valued and traded commodity.' See also the Global Water Partnership (2000), p. 36: 'The recognition of water as an economic good is central to achieving equitable allocation.'

[6] Dublin Statement on Water and Sustainable Development, Guiding Principle No. 4., www.wmo.ch/web/homs/documents/english/icwedece.html. or www.dundee.ac.uk/cepmlp/water/html/ dublin_statement.html. See also Solanes and Gonzalez-Villarreal (1999).

Past failure to recognize the economic value of water has led to wasteful and environmentally damaging uses of the resource. Managing water as an economic good is an important way of achieving efficient and equitable use, and of encouraging conservation and protection of water resources.

The formulation of the Dublin Principle emphasises that dealing with water as an economic good is a means to achieve efficient and equitable use and encourages conservation and protection of water resources. It is not an end in itself. According to Savenije and Van der Zaag: 'Water economics is about making the right choices about water resources development, conservation and allocation. Financial considerations are only a part of this "benefit-cost" analysis and seldom the main consideration.'[7] Dealing with water as an economic good in this book serves the objective of sustainable development.

Control over water is the focus of this Chapter. The rights and duties involved are in practice usually related to the use of water rather than the ownership of water.[8] In addition, the actual control over water resources through its management can significantly influence the allocation of freshwater resources, especially within states or among people. At a global level, trade and investment can in various ways be decisive in the international allocation of and control over water.

5.2   The right to use water

A right to use water at the community or national level can be acquired by: the effect of law in the case of ownership of the water resource (*ministerio legis*), appropriation (through prior use), or administrative authorisation (permits, licences or concessions).[9] According to Caponera: 'The introduction of a strict permit system and the declaration of special zones where the use of groundwater takes place represent the basic features of modern legal regimes.'[10] The renewed debate on privatisation of water resources renders importance to the issue of "ownership" of water and is therefore more elaborately discussed. At the national level, the right of a state to use water is dominated by the principle of sovereignty. The sovereignty of a state in

---

[7] Savenije and Van der Zaag (2002), p. 103.

[8] Caponera (1992), pp. 249: 'Modern legal enactments purport to limit individual and exclusive rights of ownership in favour of a centralized administrative control over groundwater through the introduction of a formal separation between the two concepts of "ownership" and "right to use."'

[9] Gupta (1996), p. 11. On the right to use water see also Caponera (1992), pp. 140-147, and Teclaff (1985), pp. 145-178.

[10] Caponera (1992), p. 248.

using its waters is qualified, for example, by the rights of its population and of other states. In case certain uses of water are granted priority, community control over water through user rights or state sovereignty could be overridden at the international level.

5.2.1   Ownership of water

In discussing water as an economic good, ownership of water is an important legal issue.[11] Legal ownership is usually protected and provides for control over that which is owned. Under the civil law system, ownership of land often implied ownership of surface as well as groundwater. In common law groundwater can generally not be privately owned. In Moslem law, groundwater is in principle considered to be a public good.[12] In Hindu law, there is no ownership of water, rather a shift to service management.[13] Roman law distinguished between private and public ownership of water, determined by the legal status of land.[14] At present, the *public trust* doctrine seems to be incorporated in constitutions of both civil and common law states. This doctrine does not easily combine with strict ownership of water. According to the *public trust* doctrine, the government must protect resources such as water that it holds in trust for the public.[15] The doctrine suggests that the ownership therefore lies with the public. In that case, a government could not sell the ownership over water. In order to sell something, one needs to own something or have permission of the owner, since you cannot transfer rights you don't have. Even in the case of legitimate representation by a government of the public, its authority to sell control over the national waters can be debated. It was, for example, stated at the Second World Water Forum that the government monopoly should not be replaced by a private monopoly and that the water resources should not be privatised, but that water resources are a common heritage and should be treated as a common property resource.[16] The question of who owns the waters of the earth so far remains. According to McCaffrey 'It would be going too far in the current state of international law to suggest that all freshwater is *res communis*. But it is critical that states begin to conceive of the *hydrologic cycle* in this way.'[17]

---

[11] Gupta (1996), pp. 4-11 and Caponera (1992), pp. 138-140.

[12] For an Islamic vision on water, including ownership and trade, see Faruqui, Biswas and Bino (2001).

[13] Gupta (2004), p. 10.

[14] Caponera (1992), Chapter 3.

[15] The doctrine can be traced back to the Institutes of Justinian (530 A.D.).

[16] See the summary report of the Forum by HRH The Prince of Orange and Rijsberman (2000), pp. 16-17.

[17] McCaffrey (2001), p. 53, who further states at p. 57:

'But a fundamental question that will have to be addressed as water resources continue to dwindle is, who "owns" the water involved? For example, is an iceberg floating in the sea beyond the limits of national jurisdiction *res nullius*, so that it is

Ownership in a strict sense implies the identification of a specific object. Although ownership of surface as well as groundwater is often referred to, it is hard to specify what is actually owned. The biophysical complexity of water makes the ownership of water in a strict sense rarely possible.[18] In its natural state, water flows and a specific water molecule will only resort in a certain territory at a given time. Ownership of water does not accompany it to the territory of another owner, indicating that the right does not concern the drops as such but the right to use water contained in a certain area at that given moment. Moreover, necessary information on the location of water is missing, especially when considering groundwater, further complicating the establishment of ownership.[19] Besides the difficulty to pinpoint the exact water supposedly owned, the amount of water is also hard to establish when water is not contained. In the case of, for example, bottled water, ownership could be more easily established.

Other objections to ownership of water *stricto senso* can be based on the vital nature of water to life. In addition, water is irreplaceable in many of its uses such as for drinking and food production. These characteristics make it undesirable to create a dependency upon a limited number of owners who control the use of water. Moreover, many cultures and religions hold water sacred and for that reason not capable of being possessed by humans.[20]

For the above reasons, reference in this book is in principle made to use of water – providing control over water to some extent – instead of ownership over water. As stated in the summary report of the Second World Water Forum: 'When we determine water rights we establish use rights – not ownership.'[21]

---

subject to appropriation? Or is it *res communis*, subject to allocation only by the international community?'.

[18] According to Dalhuisen *et al.* (2000), p. 4: 'A crucial characteristic of water from an economic point of view is that the assignment of property rights is difficult. Water falls from heaven, and flows and evaporates with no regard to any boundary, be it private, state, or national.'

[19] See Chapter two of Gleick (2000) on the water stocks and flows and international river basins in the world, where it is concluded as well that certain information is missing.

[20] According to Caponera (1992), p. 156: 'in many countries paying for water per se seems to go against religious, customary or other legal beliefs. In such cases, some countries have resorted to charging for the 'services of water supply' rather than for the water.'

[21] Summary report of the Second World Water Forum by HRH The Prince of Orange and Rijberman (2000), p. 17.

5.2.2   Sovereignty over water resources

Control over water resources at the (inter-)state level has been linked to sovereignty. A sovereign state has jurisdiction over its territory, including land, air space above the land and territorial waters. States have the sovereign right to exploit their own resources pursuant to their own environmental and developmental policies.[22] The sovereignty over natural resources provides states with the control over freshwater resources within their territory.[23] It grants states the instrument to decide the allocation of freshwater resources and enables them, for example, to implement a right of access to water or to establish user rights. Sovereignty of states is a core principle when it comes to international law, closely linked to equality of states and reciprocity, and well-established in treaty and customary law. Over time, sovereignty has not so much lost its relevance but has been subject to qualifications.[24]

The control of a state over its natural resources does not, in itself, guarantee its allocation in favour of sustainable development. Principle 1 of the ILA New Delhi Declaration well expresses the duty of states to ensure the sustainable use of natural resources.[25] The Principle includes the responsibility of states not to cause significant harm, the obligation to manage natural resources so as to contribute to their peoples' development and to the protection of the environment, the duty of states to consider the needs of future generations and a duty of all relevant actors to avoid wasteful use of natural resources.[26]

Within international water law the qualification of sovereignty is reflected in the principle of limited territorial sovereignty and limited territorial integrity.[27] The principle of limited territorial sovereignty and limited territorial integrity modifies the extreme exercise of absolute territorial sovereignty and absolute territorial integrity. The idea of absolute territorial sovereignty, as laid down in the Harmon doctrine, is basically that a state is totally free to act in whatever way it likes within its own territory.[28]

---

[22] For a thorough analysis of sovereignty over natural resources and its qualifications, see Schrijver (1997). On the issue of states trying to gain control over water, see De Villiers (2001), pp. 309-313.

[23] On territorial sovereignty over water, see Birnie and Boyle (2002), p. 301-302.

[24] Schrijver (2000).

[25] See www.un.org/ga/57/document.htm for the ILA New Delhi Declaration.

[26] Section 7.4.2 of this study elaborates on the no-harm principle, Section 7.2.2 on the right of self-determination and Section 6.3 on protection of the environment.

[27] See McCaffrey (2001), pp. 137-149, on limited territorial sovereignty.

[28] Named after the Attorney General of the United States, M. Harmon, who stated in 1895 that the US had absolute sovereignty in its use of water from the Rio Grande, a river shared between the US and Mexico. Although referred to by India in relation to the Indus as well as the Ganges, absolute territorial sovereignty was actually never applied, not even by the US. On allocation from the Rio Grande, see Lopez (1997) who concludes that the 1944 Treaty Between the United States of America and Mexico Respecting the Utilization of Waters of the Colorado and Tijuana Rivers and of the Rio

This view favours upstream states, regardless of the effect on downstream states. The counter to absolute territorial sovereignty, and equally extreme, is absolute territorial integrity, used by downstream states.[29] According to Caponera absolute territorial integrity: 'corresponds to the "theory of natural flow," whereby a state is entitled to expect that the same volume of water, uninterrupted in quantity and unimpaired in quality, flow into its territory.'[30] Within international water law the prevailing opinion is that neither absolute territorial sovereignty nor absolute territorial integrity are well-established in international law. According to McCaffrey: 'there is virtually no support, in either state practice or the writings of commentators, for the isolationist theories of absolute territorial sovereignty and absolute territorial integrity.' Although some countries may view the status of the theories differently, the concepts do appear to be eroding and seem at present to receive only little support.

The principle of limited territorial sovereignty and limited territorial integrity allows interests to be balanced as required by equitable and reasonable utilization, as well as sustainable development. Equitable and reasonable utilization can cause the interests of one state to override full exercise of sovereignty over part of the resource by another state. As stated earlier, the same principle of equitable and reasonable utilization at present includes future uses of water instead of only historical rights or prior appropriation.[31] In addition to qualified sovereignty and equitable and reasonable utilization, the emerging concept of community of interests would seem suitable for regulating water allocation toward sustainable development.[32]

5.2.3   Priority of use?

Vital human needs, sustainability and ecosystem protection were identified as subjects whose protection may require priority in the allocation of water.[33] Priority of a use means that in the case of conflict of uses or related interests, one use will receive water first to the extent that the priority is set. The relationship between different kinds of uses of international watercourses is arranged for in Article 10 of the Watercourses Convention:

> 1.   In the absence of agreement or custom to the contrary, no use of an international watercourse enjoys inherent priority over other uses.

Grande is to be amended to include the principle of equitable and reasonable utilization. See McCaffrey (2001), p. 113-128 on absolute territorial sovereignty.

[29] See McCaffrey (2001), p. 128-137 on absolute territorial integrity.

[30] Caponera (1992), p. 213.

[31] See Section 3.3.3 of this study.

[32] On community of interests, see Section 3.4.3.

[33] On priority of use, see Caponera (1992), pp. 147-148.

2. In the event of a conflict between uses of an international water-course, it shall be resolved with reference to Articles 5 to 7, with special regard being given to the requirements of vital human needs.

Within the Watercourses Convention, and within current international water law at large, no use of water is given inherent priority over another use. It is interesting to note though that according to Article 14 of the Berlin Rules states do first have to allocate waters to satisfy vital human needs in determining an equitable and reasonable use.

Under Article 10 of the Watercourses Convention, priority can be granted to a use through agreement or custom. If there is a legal recognition of the human right to water it could be argued that this would automatically imply that priority is given to basic water needs of individuals or a population. The human right to water would then provide for the agreement and/or custom that gives priority to basic human needs.[34]

In addition and in view of the special significance accorded to vital human needs, this category is likely to qualify for *de facto* priority when balancing interests in line with Articles 5 to 7.[35] According to the Statement of Understanding attached to the Convention: 'in determining "vital human needs", special attention is to be paid to providing sufficient water to sustain human life, including both drinking water and water required for production of food in order to prevent starvation'.[36] Moreover, the ILC stated in its comment to Article 7 of the Draft Watercourses Convention: 'A use which causes significant harm to human health and safety is understood to be inherently inequitable and unreasonable'.[37] In applying the principle of equitable and reasonable utilization, arguments in favour of any other use of water therefore are not likely to override the interests involved in basic access to water.

---

[34] General Comment No. 15, para. 6, states: 'priority in the allocation of water must be given to the right to water for personal and domestic uses. Priority should also be given to the water resources required to prevent starvation and disease...' See Section 4.2. of this study on the General Comment.

[35] Nollkaemper (1996), p. 61: 'A first and obvious interest that qualifies for special legal protection is the protection of vital human needs such as drinking water and household needs.'

[36] Watercourses Convention, Statements of Understanding, *I.L.M*, 36 (1997), at 719. See also Commentary to Article 10 of the Watercourses Convention, ILC Report on the work of its Forty-Sixth Session, Official Records of the General Assembly, Forty-Ninth Session, Supplement No. 10 (A/49/10), 1994, 2576. In ACUNS, *Plan of Action for Johannesburg: The Development-Environment Nexus*, distributed at PrepCom III, UN Headquarters, 28 March 2002, clean drinking water for human consumption is argued to have priority over water for agriculture and water for commercial and industrial use.

[37] ILC (1996), *Commentary to Article 7*, para. 14. In line with this Commentary, Nelissen (2002), p. 19, in relation to the use of internationally shared water resources, argues in favour of providing absolute priority to drinking water. McCaffrey (1992), p. 24, argues that human lives and health should take precedence over economic development.

Moreover, sustainability and protection of the ecosystem are to be given considerable weight. Although the status of the evolving concept of sustainability of water use remains obscure, the concept appears to at least include 'the protection of the watercourse and the protection of vital human needs, adding the need to consider long-term horizons in planning processes, and to shift from traditional notions of bilateralism towards common responsibility.'[38] The survival of ecosystems means the survival of life support systems and therefore the long term survival of all, including the human species. Sustainability without a minimal protection of the environment is simply impossible.

Nevertheless, protection of basic human needs and ecological sustainability are not sufficiently guaranteed under Article 10 of the Watercourses Convention, whose formulation – no inherent priority of vital human needs on the one hand but special regard on the other hand – renders the text somewhat ambiguous.[39] A shift in the burden of proof could provide an instrument to improve the safeguarding of these crucial elements. In cases in which vital human needs or ecological sustainability are not protected, parties would need to prove that they nevertheless have met the criteria of equitable and reasonable utilization of freshwater resources.

5.3   Management of water

According to Article 3.14 of the Berlin Rules, management of waters includes the development, use, protection, allocation, regulation, and control of waters.[40] Considering the scarcity of water, water often has to be allocated between competing uses. The effective and efficient use of water resources requires achieving set goals with a minimum of means, such as time and money. Efficient water management is hampered by, for example, water loss during transport and because of leakages. The mismanagement of water by governments has provided an important impulse for the present tendency to (re-)turn to the private management of water. It was, on the other hand, argued in Chapter 4 that access to water for basic human needs can best be complemented by community management of water.

---

[38] Nollkaemper (1996), p. 67.

[39] According to Ellen Hey (1998), p. 294: "Vital human needs' are to be included in the balancing of interests to which all uses are to be subjected, albeit with 'special regard' but not with the objective of attaining a particular result that would ensure the protection of these needs.' Gleick (2000), p. 10, on the other hand, states that: 'In interpreting Article 10, priority allocation of water in the event of conflicting demands goes to water for fundamental human needs.'

[40] See Rockström, Figuères and Tortajada (2003) on innovative approaches toward water management. Savenije and Van der Zaag view integrated water management as the foundation on which three – technical/operational, political and institutional – pillars support the sharing of international waters (the 'roof' of the temple), see Savenije and Van der Zaag (2000b) and Van der Zaag and Savenije (2000) and (1999).

Private management, community management and public management will now be further discussed.

5.3.1   Private management

Private management here refers to control by the private sector. One way of managing water is to delegate control to the private sector at the national level. A noticeable difference between the private management before the incorporation of the *public trust* doctrine and the average present form of private management is that "private" mainly used to refer to individual landowners, while "private" such as in private-public partnerships now foremost refer to multinationals or other companies, commercialising water. In most cases involving the private sector, property rights over water are not granted to private parties. The "ownership" is usually left with the state, while the private sector provides some or all services. There are many ways to transfer control over water management to the private sector. It can include retrieval of water payments as well as the management of the whole waste water sector. Contracts arranging for the transfer of control can pose conditions on the private sector but also often provide them with a long-term contract that includes profit safeguards by governments. The relative importance of private and public organizations in the water supply function differs immensely in different cases.[41] The combination of public and private sector in the management of water can be referred to as public-private partnership.[42] Considering that the waters were originally mainly controlled by the public sector, public-private partnerships involve privatisation to a certain extent.

The French corporations Vivendi and Suez are the largest water distribution companies.[43] Other players in the water industry include Saur-Bouygues, Severn-Trent, the American Bechtel-United Utilities and Degremont.[44]

In many countries, the trend toward the privatisation of publicly managed resources has now affected water management as well.[45] The reasons

[41] Dalhuisen *et al.* (2000), p. 9.

[42] WWAP (2003), pp. 380-381.

[43] Vivendi Universal is the corporation with the largest annual sales in the water industry. Vivendi is based in France and operates in more than 90 states. See Barlow and Clarke (2002), pp. 112-117. Suez, formerly Suez-Lyonnaise des Eaux, is based in France and serves about 110 million people in about 130 states. See Barlow and Clarke (2002), pp. 109-112.

[44] See Chapter 5 of Barlow and Clarke (2002) on the global water industry. Petrella (2001), pp. 68-70, presents an overview of cities whose water is managed by Lyonnaise des Eaux.

[45] Resources subjected to privatisation include mail services, telecommunications, gas and electricity, urban transport, railways, airlines, health, education and training and social security. The consequences of privatisation of branches are not all positive, as is increasingly realised due to cases such as the electricity supply problem in California.

why water is viewed these days as appropriate for the application of market mechanisms such as privatisation, include the failure of governments to manage freshwater resources effectively. Moreover, water as a common pool resource, re-enacts the tragedy of the commons, where users ignore their effect on the common pool in pursuing their own interests.[46] Privatisation is one possible response.

Furthermore, the increasing scarcity of water and improved techniques make water a potentially profitable line of business, inevitably attracting the attention of the private sector. Privatisation is often meant, through the introduction of competition, to reduce inefficiencies. For example, in California the granting of rights over water to farmers seems to have resulted in an increase in efficiency. Since 1992, farmers are allowed to sell water rights to the cities, stimulating more efficient irrigation and offering cities an alternative to the creation of new reservoirs.[47] At the international level, developing countries are regularly compelled to adopt a market approach under restructuring programmes by financial or commercial multilateral bodies such as IMF, World Bank and GATT/WTO.[48] EU law does not force any country to privatise, but it does encourage de-monopolisation through the application of competition rules and therefore stimulates movements in favour of privatisation of water management.[49] The investment required to overcome (some of) the water related problems are enormous, which partly explains the desire and the need to involve the private sector.

On the other hand, transferring control over water to the private sector entails certain risks. While a market approach in theory leads to optimal usage, the perfect market that it assumes does not exist. In practice, the market is permeated by market failures, including externalities such as health impairment or pollution not included in the price of water.[50] Moreover, even optimal use of freshwater resources could result in the allocation of water to industry over the basic needs of people or ecology. Furthermore, while publicly owned firms can be financed out of other gov-

---

On privatisation of water, see *e.g.*, Gleick, Burns, Chalecki, Cohen, Cushing, Mann, Reyes, Wolff and Wong (2002), Chapter 3.

[46] See Dalhuisen *et al.* (2000), p. 4. See also Murty, James and Misra (1999), p.107, on the management of common pool resources under various property rights regimes.

[47] Whether the resulting allocation of water rights is to be viewed as positive or not is debatable due to factors such as possible speculation with water rights, see Barlow and Clarke (2002), p. 73.

[48] See Section 5.4 of this study.

[49] Hancher (1997), p. 285.

[50] According to Savenije and Van der Zaag (2002), p. 104: 'Within sectors, water markets and marginal cost pricing may in some cases be compatible with the concept of Integrated Water Resources Management, provided all externalities are indeed "internalized" and transactions are regulated by a public body'. The authors continue to argue that 'for the allocation of water between sectors no markets are required nor are these desirable.' See Section 8.4 of this study and Annex II on the pricing of water.

ernment sources if necessary, the privatised part of the water industry needs to make a profit in order to function.[51] Shareholders often require profit to increase. Although in the long-term this is bound to lead to failure, in the short term the profit made by raising prices might be paid to the shareholders. The result is that the earnings may not so much be invested in, for example, a better water infrastructure. In addition, accountability may be diminished in case of involvement of the private sector that might have to answer to and share information with its shareholders instead of the public. Also, private control over water is hard to combine with the *public trust* doctrine.

Furthermore, in the case of converting a system of private ownership (back) to public ownership, financial compensation can be expected to be required if expropriation or breach of contract is involved. The need for compensation in case of expropriation is acknowledged under international law. The remaining discussion mainly focusses on the determination of what is "appropriate" compensation.[52] In case compensation results in large amounts taking into account the large profits that multinationals often make compared to the national incomes of developing countries, developing countries in particular might not be able to meet the legal requirements to regain control over their natural resources.[53] On the one hand, compensation is not to result in such a *de facto* control. On the other hand, the compensation is to be sufficient not to disproportionally harm the interests of investors and, moreover, to not create a bad climate for investment.

Another risk entailed by the transfer of control over water resources to the private sector is that the level of investment required means that the supply-chain of water in its entirety tends to become a natural monopoly, calling for public provision of at least the distribution network.[54] Moreover, the quality of water has to be guaranteed necessitating strict control, especially if the water is transported through one network by different users, implying a risk of free-rider behaviour.[55] A question raised by privatisation is whether or not one can prevent foreign take-over if water ownership or control is privatised and the principles of free trade are applied,

---

[51] Dalhuisen *et al.* (2000), p. 9.

[52] Schrijver (1997), pp. 350-359, addressed also the standard of compensation such as 'appropriate' compensation in the 1962 Declaration on Permanent Sovereignty over Natural Resources.

[53] According to Barlow and Clarke (2002), pp. 83-84, there are about 45,000 transnational corporations and the combined annual sales of the largest 200 are estimated to be greater than the economies of 182 states all together.

[54] Dalhuisen *et al.* (2000), p. 8. See also Hunter, Salzman and Zaelke (2002), p. 814: 'In practice, however, privatization of water supplies has frequently led to monopoly control of water resources (with their own problems of inefficiency and inequity), as well as to reduced services for the poor and lower ecosystem protection.'

[55] Dalhuisen *et al.* (2000), p. 14.

once water is viewed as a tradable good.[56] An additional reason to reconsider the arguments in favour of privatisation, is that the private sector is not necessarily efficient in providing water to people.[57] The position and power of the water industry has to be inventoried, and its promises as well as its dangers thoroughly identified.[58]

Inefficiency in case of privatisation is well illustrated by the following examples concerning the UK and France. In the UK under Thatcher, water management was privatised.[59] The regimes vary for England, Scotland and Wales. Privatisation in England includes the rare case that "ownership" of water facilities has actually been sold.[60] In England the price of water has risen since privatisation in 1989, causing the government to impose a special windfall tax on excess profits in 1997.[61] The provision of services and the number of leakages seem not to have improved; the price rise cannot therefore be explained in terms of efficiency. An example of the more common form of privatisation, *i.e.* public-private partnership, is presented by France, where water management, operation and collection of revenues are transferred from the government to corporations through concessions or leases.[62] The privatisation in France has resulted in large increases in prices of water over the last couple of years, raising the profit levels for private companies.[63]

The rise in prices such as in France as well as in the United Kingdom, where water is not even particularly scarce, raises concern over the affordability of water for people in poorer countries where water may be privatised. For example, in 1997 the responsibility for the water supply in Manila was handed over to two private-enterprise groups - one consisting of a Philippines company, the American Bechtel corporation and the British United Utilities, and the other consisting of one Philippines company

---

[56] Hancher (1997), pp. 288-289, raises a case in the UK in which a bid of Lyonnaise des Eaux SA was not accepted and a take-over was stopped by the British authorities and agreed upon by the EC Commission to safeguard continued effective regulation.

[57] According to Hunter *et al.* (2002), p. 814: 'In practice, however, privatization of water supplies has frequently led to monopoly control of water resources (with their own problems of inefficiency and inequity), as well as to reduced services for the poor and lower ecosystem protection.' They continue to state conditions in case of privatisation of water resources.

[58] On risks and benefits of privatisation, see Gleick, Wolff, Chalecki and Reyes (2002), who, at p. 43, 'strongly recommend that any efforts to privatize or commodify water be accompanied by formal guarantees to respect certain principles and support specific social objectives.'

[59] Vass (2002).

[60] Barlow and Clarke (2002), p. 89.

[61] Petrella (2001), pp. 74-75.

[62] Barlow and Clarke (2002), p. 89. At the same page they also mention a third form of privatisation: transfer through contract of management from government to corporations for an administrative fee without collection of revenues. According to the 1992 French Water Act, Law No. 92-3 of 3 January 1992, the use of water is a public matter.

[63] Petrella (2001), p. 73, referring to the parliamentary report of Guelle.

and the French Lyonnaise des Eaux. The group in the richer part of Manila is said to charge less than the one in the poorer area.[64] Such a situation is hard to square with the principle of equity.

Cases involving both government and company failure have resulted in disputes.[65] Some of those disputes are referred to the International Centre for Settlement of Investment Disputes (ICSID) of the World Bank, such as the conflict between Vivendi and Argentina, and the pending *Cochabamba* case (Bechtel *vs.* Bolivia) that resulted from the departure of Bechtel due to massive protests of people against water privatisation and the raise of prices.[66] In order to receive a loan for water services, in 1998 the World Bank required Cochabamba to privatise its public water utility. Control over the water utility was transferred to Aguas del Tunari, a subsidiary of Bechtel. The protest of the people of Cochabamba resulted in the government of Bolivia to cancel its contract with Bechtel. Bechtel in this case invokes a Bilateral Investment Treaty (BIT) between Bolivia and The Netherlands to sue the Bolivian government before ICSID where it claims US$40 million expropriation rights.[67]

5.3.2   Community management

Community management provides people with a certain control over their water resources and can take many forms.[68] There are strong indications that the World Bank may come to prefer this approach. The International Trade Unions, as represented by Public Services International (PSI), reject privatisation of water and sanitation services, on the basis of the belief that these services should be owned and managed by democratic and accountable public bodies close to the communities.[69]

Pre-colonial India seems to provide an example of successful community management, although given the caste system the definition of community is open to question.[70] However, its population these days - and therefore the number of (potential) conflicts over water - is much higher

---

[64] See Petrella (2001), pp. 9-10, also for other examples of major cities in developing countries of which the control over (part of) water has been transferred to private companies: Mexico City, Hanoi, Buenos Aires, Casablanca and Moscow.

[65] For examples of national court cases, see Barlow and Clarke (2002), p. 116, p. 122, and p. 125, 190-191, including cases on prices, loss of jobs, pollution, corruption, and pumping for bottled water.

[66] See www.worldbank.org/icsid. Vivendi vs. the government of Argentina, was based on a BIT between France and Argentina, and related to a water contract with the City of Tucumán. See also Barlow and Clarke (2002), pp. 177-178.

[67] See Barlow and Clarke (2002), pp. 91, 154-156 and 177. See also www.worldbank.org/icsid/cases/pending.htm and www.Bechtel.com.

[68] Schuttelar, Ozbilen, Ikeda, Hua, Guerquin and Ahmed (2003).

[69] See Statement by Public Services International on behalf of International Trade Unions, World Water Council (2000), p. 92.

[70] See Petrella (2001), p. 15, where he refers to the pre-colonial situation in India as a positive example of destatization.

and this complicates comparison. Nevertheless, the present day Rajiv Gandhi Watershed Mission seems to have been quite successful.[71] Another form of community ownership is through the creation of Water Users Associations (WUAs) to which the management of water is transferred. The WUAs aim at: 'optimum utilization of available water through a participatory process that endows farmers with a major role in the management decisions over water in their hydraulic unit.'[72] The co-management of watersheds by American Indians in the USA provides an example of the relevance of community management of freshwater resources in developed countries.[73] After Bechtel left Cochabamba, the management of the local water company (SEMAPO) was handed over by the Bolivian government to the community.[74] In South Africa, the Okavango Liaison Group presents an example of organised local communities at the catchment basin level.[75] The Farmer Managed Irrigation Systems (FMISs) in Nepal present an example of local water management for the benefit of a whole community.[76]

From a sustainable development viewpoint, there is much to be said in favour of community management. Placing responsibility with the community can be argued to enable participation, to facilitate the implementation of access to water for all people, and to increase awareness of water issues. Community management is likely to be most promising if all interest groups are well represented. The WSSD Plan of Implementation, for example, addresses the promotion of women's equal access to and full participation in decision-making at all levels, and the improvement of their status through full and equal access to land.[77]

Especially in cases such as government failure, as well as in case of specific interests or knowledge of indigenous or tribal people, community control and management over freshwater resources offers an interesting alternative.[78] Community management can also offer an alternative to private sector involvement. Within a community-public partnership the compatibility of water management with requirements of sustainable development could be supervised and, for example, more easily take into account factors such as long-term interests outside the area.

---

[71] See Section 8.4 of this study.
[72] Salman (1997), p. 1.
[73] See Goodman (2000) on Indian tribal co-management, underlining the importance of their traditional ecological knowledge and technical expertise. On water management by Indian tribes in the US, see Kannler (2002).
[74] Barlow and Clarke (2002), pp. 186-187.
[75] Barlow and Clarke (2002), p. 199.
[76] Barlow and Clarke (2002), p. 235.
[77] WSSD Plan of Implementation para. 6(d).
[78] According to McCaffrey (2001), p. 172: 'the concept of community management can be taken further, and indeed it may have to be in the twenty-first century as the per capita availability of potable water continues to dwindle.'

5.3.3   Public management

Public management remains an important and widespread way of water management. In addition to the advantages and disadvantages of public management of water resources that can be deduced from the arguments in favour and against private ownership, national security provides a further argument in favour of public control over water. This does not necessarily refer to situations of war, but concerns the survival of the state in a broader sense. In practice, freshwater resources are often viewed as a matter of national security. The link between control over water and national security partly explains the concerns of Arab countries over Turkish control over the Euphrates and their rejection of the offer of Turkey in 1987 to export its waters (on a commercial basis) by jointly building a "peace aqueduct".[79] Self-sufficiency, at a minimum, reduces vulnerability and increases national security. In addition, water can be used as an instrument of war. For example, diversion of the Jordan river by Syria led to Israeli air strikes in 1965 (B and C p. 72) and today again causes tensions. In 1974 Iraq threatened to bomb the Tabga dam in Syria. Turkey threatened to reduce the water supply downstream in order to stop Syrian support for the Kurds in south-east Turkey.[80]

A downside of the recognition of water resources as a matter of national security is that it can hamper the exchange of information. It can thereby complicate the cooperation required for sustainable development of freshwater resources. Another negative effect of the national security approach is that it has resulted in overprotection of the agricultural sector, user number one when it comes to water. For example, the large subsidies in this sector in Europe stimulate unsustainable production and arguably affected the interests of developing countries.

Mismanagement by the public sector has been one of the main reasons for a renewed flight toward the private sector. But privatisation, to whatever extent, actually requires a strong government in order to regulate and control it and to enforce social and ecological conditions.[81] A consensus seems to exist that governments should not sell "ownership" over water. In case the private sector is involved, the social and ecological interests are to be safeguarded by the government. Involvement of the community not only poses a condition to provide space for participation, it also provides an alternative to private or public management. Any contracts with the private sector or other parties that transfer control over water are to be for a limited period of time only and need to establish certain conditions. For example, the affordability of water for the poor and water for ecosystems

---

[79] McCaffrey (2001), p. 284.
[80] See Petrella (2001), p. 45.
[81] Hunter *et al.* (2002), p. 814: 'In short, privatizing water resources can improve efficiency, but, unless significantly regulated and shaped, privatization can also undermine environmental and equity values.'

are to be guaranteed, the quality of water is to remain under public supervision and accountability is to be safeguarded.

Public management is furthermore required in dealing with the use of water at the international level and between states. Equitable and reasonable utilization is the main principle of international water law, dealing with the allocation of water between states. In the determination of such utilization, the actual uses of water are taken into account when weighing the different factors involved, including those at the national and community level, emphasing the interaction between the policy-levels.

5.4   A supportive and open international economic system

According to, for example, the Rio Declaration and the UNFCCC, states are to aim for a supportive and open international economic system. Rio Principle 12 states: 'States should cooperate to promote a supportive and open international economic system that would lead to economic growth and sustainable development in all countries, to better address the problems of environmental degradation.' The objective is categorised as a principle under Article 3.5 of the UNFCCC. Means to achieve such an economic system are international trade and investment.[82]

Treating water as an economic good between states can result in international trade in water as well as in international investment in water. International trade in water can have a large impact on the use and allocation of water depending on whether it concerns a transfer of user rights or if water itself is regarded to be a tradable good or commodity. As soon as this is regarded as trade in water under the trade regimes of the World Trade Organization (WTO) and/or the North American Free Trade Agreement (NAFTA) the consequences could be enormous. Considering the major impact they could have, the WTO and NAFTA will be separately discussed. Investments in water or, even more, investments requiring a certain management of water, can also be of decisive influence upon the use of water. The Bretton Woods institutions play an important role in this respect.[83] Although international trade and investment, as part of international economic law, are often dealt with as autonomous fields of law, their impact on water management should be viewed within the criteria set by interna-

---

[82] On international trade, see Verbruggen (1999).

[83] The Bretton Woods institutions, as well as the changes that took place within them during the years, have to be viewed in a historical context. In July 1944, the UN Monetary and Financial Conference was held in Bretton Woods, New Hampshire, USA, resulting in the creation of the International Monetary Fund and the World Bank. The International Trade Organization never came into existence, although a charter (Havana Charter) was created. The three institutions were supposed to work in a complementary fashion. The main goals of the creation of these institutions was to increase worldwide well-being and to prevent the sort of economic discrimination that was seen as partly paving the way for World War II.

tional water law and other relevant fields of law such as the body of human rights law. The actual impact of trade, of the WTO and NAFTA trade regimes and of investment in water is analysed below to see if and in what way they can be compatible with sustainable development.

5.4.1   International trade in water

Throughout history, humans have transported water.[84] The transfer of water is, for example, regulated in a treaty between Lesotho and South Africa.[85] An example of the transfer of water between states is the tender between Turkey and Israel.[86] Water transfer also takes place in the larger Southern African Development Community (SADC) region.[87] Between Mexico and the United States an informal water market seems to have come into existence.[88] Trade in water can also be identified within states. For example, Hong Kong pays mainland China for the delivery of drinking water.[89] A future possible aspect of trade could be trade in icebergs.[90] Exchange of water does not, however, necessarily constitute trade in water, but depends arguably on such elements as whether the payment is for the actual water or for the delivery of the water, in which case it constitutes a trade in services.[91] If it is the service that is paid for and water itself is not viewed as a product, for example, its trade and the absence of strict ownership can pose less of a problem also for cultures in which water is thought to be sacred.

---

[84] See McCaffrey (2001), pp. 8-15 on water transfers, further elaborating upon the 'Great Man-Made River' in Libya, diversions in California, the hydraulic undertakings in the American West, and the role of anthropocentrism.

[85] Treaty on the Lesotho Highlands Water Project, 24 October 1986, between the Kingdom of Lesotho and the Republic of South Africa. Boadu (1998).

[86] In May 2001 the special tenders committee of the government of Israel has issued the first tender for importing water from Turkey (35-50 million cubic meters of water annually for 5-10 years), see Haaretz special for the on-line edition, www2.haaretz.co.il/special/water-e/d/364915.asp. On August 6 2002, Israel agreed to buy 50 million cubic meters of water from Turkey for the next 20 years, see Associated Press, Wednesday, August 7 2002, http://enn.com/news/wire-stories/2002/08/08072002/ap_48059.asp.

[87] See Heyns (2002), concluding, at p. 175 that: 'Experience in SADC clearly show that interbasin water transfer schemes can be an effective tool to enhance joint regional co-operation, and improve water resource management.'

[88] Sánchez (1997), pp. 275-276. On water markets, see also Mariño and Kemper (1999).

[89] See www.planetark.org/dailynewsstory, 'Greenpeace says metals pollute HK drinking water' (article, China, 18 December 2000): 'Hong Kong pays mainland China HK$2 billion (US$256 million) a year for drinking water.'

[90] Geon (1997) concludes at p. 301 that international law on iceberg appropriation is unclear.

[91] Gleick, Burns et al. (2002), Chapter 2.

Water can, for example, be more sustainably traded not as a commodity in itself but indirectly by import and export of water-intensive products produced in water rich areas and imported by arid regions, *i.e.* by trade in virtual water.[92] In a sense, it can be argued that trade in water already exists since all kinds of products, especially drinks, mainly consist of water. This is called the virtual water trade.[93] As a separate product, bottled water concerns a certain amount of fresh water and is therefore more easily recognised as a tradable commodity. An increasing demand and diversification in the bottled water industry now make it a promising (international) market.[94] The number one in water bottling is Nestlé (encompassing 68 brands, including Perrier, Vittel, Volvet, and San Pellegrino), followed by players such as Danone (including Evian, Volvic, Frarrarelle, Cannon, Villa del Sur), Coca Cola and Pepsi Cola. The image of bottled water being healthier than tap water appears to be largely unfounded.[95] Almost half of the bottled water is consumed by Western Europeans – on average 85 litres per person per year – who often live in countries with safe tap water.[96]

Despite the scarcity of water in many areas and the quantitatively and qualitatively unequal distribution of water, water in itself has not been favoured as a trading commodity until recently.[97] Reasons for not regarding water as a tradable commodity include the difficulty of transport because of its voluminous nature (bulk commodity) and the difficulty to specify the volume and quality unless contained and therefore to identify the exact commodity that is traded.[98] Moreover, until recently, the dominant conviction was that water is a public asset managed through government.[99]

Dealing with water as a tradable commodity has only relatively recently gained support, mainly within the private sector and by governments and intergovernmental organizations. Reasons for water becoming a tradable commodity include improved technology to transport bulk water, the large investments by states and intergovernmental organizations in water management, a change in state regulation and controls enabling a profitable water sector, and a different outlook on water altogether.[100] Moreover,

---

[92] The term "virtual water" was first defined by J.A. Allan as water embedded in commodities, see www.wateryear2003.org.

[93] See on virtual water trade, *e.g.*, Hoekstra (2003).

[94] Barlow and Clarke (2002), pp. 142-145, Petrella (2001), pp. 80-82.

[95] See Barlow and Clarke (2002), p. 143, referring to a 1997 FAO study.

[96] For facts and figures on bottled water, see www.wateryear2003.org.

[97] On water and trade, see De Haan (1997), in favour of valuing water economically although aware of (environmental) incompatibilities between trade and sustainable development.

[98] De Haan (1997), pp. 246-247.

[99] Since society appears to regard water so essential that it must be made available to all regardless of their ability or willingness to pay the market price for it, it can be classified as a "merit goods", better produced by the state.

[100] See, *e.g.*, De Villiers (2001), pp. 275-281, on Medusa bags and other ways of transporting water.

water will probably become more expensive considering the increased scarcity of water. A higher price of water, combined with the fact that a certain demand is guaranteed since people need water, enables cost-recovery and profit.

A large part of the international community, primarily communities and NGOs, is still very much against viewing water as a tradable commodity.[101] One of the main concerns about putting water up for sale relates to the question of who will buy it for the poor and for the environment. Moreover, trade in water is a hot issue in water-rich countries such as Canada.[102] In Canada, many fear not only that large-scale diversion of water will devastate ecosystems but also that closing one contract on trade will make it impossible to close the tap because of the NAFTA and WTO regulations.[103] One question that trade in water will raise concerns the balancing of interests between importing and exporting states in such cases where, for example, a contract results in the over-exploitation of water resources degrading the environment of the exporting country.[104] Under certain circumstances, however, the international export of water could provide for the distribution of water in a sustainable manner.[105] Specifically in cases of water-stress emergencies, temporary water assistance is likely to be compatible with sustainable development.[106]

5.4.2   The WTO and NAFTA

The WTO as well as the NAFTA will have a major role in deciding whether or not and to what extent water will be dealt with as a tradable commodity and whether or not the provisions allow states to take protec-

---

[101] See, *e.g.*, Barlow and Clarke (2002), Chapter 8. At p. 207 they argue that: 'The move to commodify depleting global water supplies is wrong – ethically, environmentally, and socially. It ensures that decisions regarding the allocation of water center almost exclusively on commercial, not environmental or social, considerations.'

[102] Barlow and Clarke (2002), p. 136 and pp. 192-193.

[103] De Haan (1997), pp. 249-251, on plans on trade of water between British Columbia and the USA, in which case the government of British Columbia opposes the diversion plans but might be forced to accept them under NAFTA regulation, prohibiting export restrictions. See Hunter *et al.* (2002), p. 815, for an elaboration on the case of the Great Lakes, concerning a 1998 proposal from a Canadian company to export 158 million gallons from Lake Superior to Asia. For the discussion taking place in Canada concerning water export, see also De Villiers (2001), pp. 246-254; Little (1996); and Baumann (2001).

[104] De Haan (1997).

[105] On import of water and other means to increase the amount of available water, see De Villiers (2001), pp. 284-293.

[106] According to De Waart (1997), p. 118, future trade agreements relating to clean fresh water should be applied in the context of the right of a specific access to clear fresh water. He continues by stating that if international trade cannot realise this right, it should be effected through international migration.

tive measures for social or environmental reasons.[107] Once water is re-garded as a tradable commodity under the WTO regime, the latter's dis-pute settlement panel can force a country to restrict its protective legisla-tion. The way water is regarded under those regimes can be of great impact to the use of water. Therefore, the question of compatibility of the regimes with sustainable development is vitally important.

From 1948 until the establishment of the WTO in 1995, the General Agreement on Tariffs and Trade (GATT) was the principal agreement regu-lating trade between states.[108] The main aim of the GATT and its related side agreements, referred to as the GATT system, has been to promote the liberalisation of trade. The aim of liberalisation has led to a policy of first reducing, and ultimately eliminating trade restrictions. The GATT system has succeeded in significantly reducing tariffs. With the entry into force of the Uruguay Round agreements on January 1, 1995, the WTO was estab-lished.[109] The GATT has now been integrated into the WTO. The GATT / WTO is mainly about commerce. It does acknowledge the special position of developing countries, especially through the principle of non-reciprocal treatment.[110] The objective of the WTO, as stated in its preamble, includes compliance of the use of resources with the objective of sustainable devel-opment and, moreover, calls for special attention to the economic devel-opment of developing countries.[111]

As yet, the status of water within WTO law remains uncertain. Under GATT, water would most probably fall within the definition of good, but it could also be a service. The tariff list contains bottled water as well as natu-ral water.[112] Article XI GATT deals with the elimination of quantitative restrictions. It prohibits measures other than duties, taxes, or other charges. In the case of critical shortages, the exporting party can under Article XI(2)(a) temporarily exclude water from this prohibition.[113] Moreover, Article XX GATT contains general exceptions. The adoption or enforce-ment of measures taken under this article is not prevented by GATT. Arti-cle XX(b) relates to measures necessary to protect human, animal or plant

---

[107] Hunter *et al.* (2002), p. 814-815: 'NAFTA and other trade agreements thus might prevent countries from putting export controls on water supplies, thereby limiting their ability to ensure water benefits to local populations.'

[108] For an elaboration on the WTO/GATT see Pescatore, Davey and Lowenfeld (1995).

[109] Agreement Establishing the World Trade Organization, Marrakesh, 15 April 1994, entry into force: 1 January 1995, WTO *Legal Texts*, 3, and 33 *ILM* (1994), 13. The WTO completed the Bretton Woods construction based on three complementing bod-ies.

[110] See Article XXXVI, para. 8 GATT. See also Article XVIII GATT. On the principle of non-reciprocal treatment, see Section 7.2.3 of this study.

[111] Davey (1995), p. 11.

[112] For the tariff classification waters are mentioned, including natural or artificial wa-ters, aerated waters, not sweetened or flavoured, as well as ice and snow.

[113] Article XI(2)(a): 'Export prohibitions or restrictions temporarily applied to prevent or relieve critical shortages of foodstuffs or other products essential to the exporting contracting party'.

life or health. Article XX(g) deals with measures relating 'to the conservation of exhaustible natural resources if such measures are made effective in conjunction with restrictions on domestic production or consumption'. In order to fulfil the requirements of Article XX, measures must not be arbitrary or unjustifiable discriminatory, or a disguised restriction.

The WTO dispute settlement system includes decisions of the Dispute Settlement Body (DSB) and can arrange for compensation or even retaliation.[114] The decision of the Panel in the *Tuna-Dolphin* cases (Mexico *vs.* US) in the early nineties, in which it was decided that the USA could not unilaterally – and with extra-territorial effect – ban imports of tuna from other countries that might not have taken Dolphin protective measures when catching tuna, caused great commotion among environmental groups and is generally regarded as one of the elements that triggered much protest.[115] In response to this pressure, in 1992 GATT issued a report on trade and environment.[116] It also reactivated its Committee on Environmental Measures and International Trade, and a WTO committee on trade and environment was established with the aim of making trade and environmental protection mutually supportive.[117] In the *Shrimp-Turtles* case (India, Malaysia, Pakistan and Thailand *vs.* US) the prohibition by the US to import shrimps from countries not using certain turtle excluding devices was found by the Appellate Body to constitute unjustifiable discrimination and reference was also made to Principle 12 of the Rio Declaration on Environment and Development: 'Unilateral actions to deal with environmental challenges outside the jurisdiction of the importing country should be avoided. Environmental measures addressing transboundary or global environmental problems should, as far as possible, be based on international consensus.'[118] However, in the *Shrimp-Turtles* case the Appellate Body took a less restrictive approach toward the exceptions in Article XX(b) and (g) GATT than in the *Tuna-Dolphin* cases and the decision provides better opportunities to take into account environmental considerations. The Body referred to 'contemporary concerns of the community of nations about the protection and conservation of the environment' and to the aim of sustainable development in the Preamble of the Agreement Establishing the WTO. On the one hand, it would appear that the WTO is wary of allowing such considerations into its system. On the other hand, the cases do discourage

---

[114] On water and WTO dispute settlement, see Girouard (2003) and Davey (1995), pp. 70-80. On the *Tuna-Dolphin* cases and the *Shrimp-Turtles* case see also Kentin (2001), pp. 82-85, and De Haan (1997), pp. 258-259.

[115] See *Tuna-Dolphin I* case, Panel Report on United States - Restrictions on Imports of Tuna, *GATT Doc.* DS21/R, BISD 39S/155, and *Tuna-Dolphin II* case, Panel Report on United States - Restrictions on Imports of Tuna, *GATT Doc.* DS29/R.

[116] See GATT Secretariat (1992), *Trade and the Environment*, reprinted in: *International Trade 90-91*, vol. 1 (1992), pp. 19-47.

[117] See Davey (1995), pp. 84-85.

[118] Report of the Appellate Body on United States-Import Prohibition on Certain Shrimp and Shrimp Products, WT/DS58/AB/R, 12 October 1998.

disguised restriction based on environmental arguments that could result in discrimination of developing countries. On the basis of the *Tuna-Dolphin* cases and the *Shrimp-Turtles* case, it can be concluded that multilateral measures that provide for environmental protection, under certain conditions, could very well be accepted under GATT.

The WTO's General Agreement on Trade in Services (GATS) regime was established in 1994 and lists many water services, such as wastewater treatment and irrigation.[119] New sectors are added to it by means of ongoing negotiations. The 2001 EU proposal to add a section on Trade and Environment includes 'the reduction, or, as appropriate, elimination of tariff and non-tariff barriers to environmental goods and services' and could further limit the possibilities to protect social and ecological interests in water.[120] On the other hand and according to the WTO:[121]

> The number of Members which have so far made GATS commitments on water distribution is zero. If such commitments were made they would not affect the right of Governments to set levels of quality, safety, price or any other policy objectives as they see fit, and the same regulations would apply to foreign suppliers as to nationals. A foreign supplier which failed to respect the terms of its contract or any other regulation would be subject to the same sanctions under national law as a national company, including termination of the contract. If termination of a contract were involved, the existence of a GATS market-access commitment would be irrelevant. A GATS commitment provides no shelter from national law to an offending supplier. It is of course inconceivable that any Government would agree to surrender the right to regulate water supplies, and WTO Members have not done so.

NAFTA is an agreement between the United States, Canada, and Mexico that aims at removing barriers to trade and investment among those countries.[122] NAFTA entered into force on 1 January 1994. NAFTA builds on the rights and obligations under the GATT. According to its Preamble, the purposes of NAFTA are to be consistent with environmental protection and conservation. The preamble states that the parties resolved to: 'preserve their flexibility to safeguard the public welfare; promote sustainable development; strengthen the development and enforcement of environ-

---

[119] Barlow and Clarke (2002), pp. 167-170.

[120] Fourth Ministerial meeting of the WTO, Qatar, November 2001.

[121] Quoted from the WTO website as on 18 August 2003,www.wto.org/english/ tratop_e/serv_e/gats_factfiction8_e.htm.

[122] North American Free Trade Agreement (NAFTA), Washington, Ottawa and Mexico City, 17 December 1992, entry into force: 1 January 1994, 32 *ILM* (1993), 289 and 605. See www.nafta-sec-alena.org. Negotiations are taking place to establish a Free Trade Area of the Americas (FTAA), based upon NAFTA and Mercosur, see Barlow and Clarke (2002), pp. 170-176.

mental laws and regulations...' One of the main objectives of NAFTA mentioned in Article 102(1)(a) is to eliminate barriers to trade. According to Article 103 NAFTA, the rights and obligations under GATT and other agreements are affirmed. Article 104 NAFTA regulates the relation to environmental and conservation agreements. Under Article 104, specific trade obligations of certain international environmental agreements can in certain circumstances take precedence over the provisions of NAFTA.[123] Article 1110 NAFTA regulates expropriation and compensation of foreign investment. The article prohibits the nationalisation or expropriation of an investment of another NAFTA party, but allows for it under cumulative conditions including the case where it concerns a public purpose and the stipulation that compensation equivalent to the fair market value of the investment is paid. Article 1114 NAFTA deals with environmental measures. Article 1114(1) NAFTA states:[124]

> Nothing in this Chapter shall be construed to prevent a Party from adopting, maintaining or enforcing any measure otherwise consistent with this Chapter that it considers appropriate to ensure that investment activity in its territory is undertaken in a manner sensitive to environmental concerns.

The North American Agreement on Environmental Cooperation (NAAEC) also deserves note.[125] As the environmental side agreement to the NAFTA, it could influence the application of NAFTA in favour of environmental considerations. Article 1 NAAEC includes the objectives to:

> (a) foster the protection and improvement of the environment in the territories of the Parties for the well-being of present and future generations; (b) promote sustainable development based on cooperation and mutually supportive environmental and economic policies... (d) support the environmental goals and objectives of the NAFTA...

Water and trade have been discussed under the NAFTA. NAFTA does not exclude water by its definitions and like the WTO has mentioned water

---

[123] See Davey (1994), p. 85. Article 104(1) NAFTA refers to the 1973 Convention on International Trade in Endangered Species of Wild Fauna and Flora (as amended 22 June 1979), the 1987 Montreal Protocol on Substances that Deplete the Ozone Layer (as amended 29 June 1990), the Basel Convention on the Control of Transboundary Movements of Hazardous Wastes and Their Disposal (on its entry into force for the NAFTA parties), or agreements set out in Annex 104.1.

[124] Cases related to this issue are The Ethyl Corporation case and The Metalclad case, see Kentin (2001), pp. 86-89.

[125] North American Agreement on Environmental Cooperation Between The Government of the United States of America, The Government of Canada, and The Government of the United Mexican States, 13 September 1993, entry into force 1 January 1994, 32 I.L.M. 1480. See www.cec.org. See also Saunders (1994).

in its tariff lists. The parties to NAFTA, however, have declared that no rights to the natural water resources of any party to NAFTA are created by NAFTA and that water is only covered by trade agreements if it has entered into commerce and become a good or product.[126] The statement goes on to say that parties are not obliged to exploit their water for commercial use or to start exporting water in any form. Furthermore, it is declared that:

> Water in its natural state in lakes, rivers, reservoirs, aquifers, water-basins and the like is not a good or a product, is not traded, and therefore is not and never has been subject to the terms of any trade agreement.

5.4.3   Investment in water

Major investments are required to meet the demands for water and to fight its degradation. The projects involved can vary widely from flood protection, dams, and water quality improvement to capacity building.[127] Investments are also needed to create and implement legislation on water. In view of the various investors and the effect of their investments, the principle of cooperation and coordination are of great importance.[128] Investments can furthermore impose conditions on water management. Yet investments in themselves do not guarantee a contribution to the sustainable development of freshwater resources. The failure to address social and ecological interests adequately as well as economic interests and the failure to involve groups other than OECD states can be seen as one of the main reasons for the collapse of the Multilateral Agreement on Investment (MAI).[129]

Among the regimes that specialise in water, and are able to invest in its management, are the European Union (EU), the Food and Agriculture Organization (FAO), United Nations Development Programme (UNDP), United Nations Environment Programme (UNEP), UN Conference on Trade and Development (UNCTAD) and the Asian Development Bank.[130]

---

[126] Statement by the governments of Canada, Mexico and the United States of 1993. Especially in Canada export of water poses a sensitive subject, see for example Little (1996).

[127] On capacity building investments, see Global Water Partnership (2000), pp. 45-47, and idem, pp. 75-84, on investing in water.

[128] On the need for donor coordination, see Boisson de Chazournes (1998), p. 75.

[129] See Schrijver (2000), p. 86. A strict trade or financial approach that appears to be ignoring other interests involved raises more and more opposition, witnessing the failure in October 1998 to come to a multilateral agreement on investments (MAI) and of the WTO Millenium Round in Seattle in December 1999.

[130] See Petrella (2001), Appendix, pp. 124-125, for a list of international organisations and UN bodies specifically involved in water-related issues.

As to the principle ones, the Organization for Economic Cooperation and Development (OECD) has committed itself to support water security and to contribute to the effort to reconcile economic, environmental and social policy objectives related to water in the context of sustainable development through various means such as collecting water resource data and developing pricing strategies that properly incorporate social objectives.[131]

The Global Environment Facility (GEF) was established in 1991 and restructured in 1994 and operates through the UNDP, UNEP and the World Bank.[132] The GEF finances actions addressing six threats to the global environment: the loss of biodiversity, climate change, the degradation of international waters, ozone depletion, land degradation and persistent organic pollutants (POPs). It has assisted in more than 100 water related projects in various countries. The GEF has promised to scale up its contributions toward solving water and land degradation problems and toward increasing public and private participation in addressing the global water crisis by: increasing financial support to water and related land resources activities; catalysing public and private investments, requiring better pricing; supporting regionally integrated land and water resources activities and the implementation of integrated ecosystem management efforts; and protecting water resources from land-based pollution.[133] Its resources however are not sufficient to meet the needs and finance such actions.[134]

Investment through international organizations such as the Bretton Woods institutions, influences the management of water in various countries. One of the main purposes of the International Monetary Fund (IMF) is 'to facilitate the expansion and balanced growth of international trade' (Article I(ii)).[135] The IMF is mainly dealing with financial issues such as international currency stability by means of assisting its member states in correcting their balance of payments deficits and by promoting or requiring certain standards of international monetary conduct. One of the criteria for the granting of loans is whether it will stimulate development of the private sector and the private production of goods and services traditionally operated by the state or cooperatives.[136] The IMF policy of encouraging programs of restructuring toward liberalisation (decentralisation, reducing

---

[131] See pledge of OECD, World Water Council (2000), pp. 72-74.

[132] See www.gefweb.org.

[133] See pledge by the Global Environment Facility, World Water Council (2000), p. 72. According to its website, GEF allocated almost $360 million to international waters initiatives over the period 1991 to 1999. See also www.gefweb.org/meetings/ WaterForum.

[134] According to the GEF website: 'In August 2002, 32 donor nations pledged nearly $3 billion to fund the work of the GEF for the next four years.'

[135] Chatterjee (1992a). The IMF came into existence on 27 December 1945, following the consideration of its Articles of Agreement at the Bretton Woods conference.

[136] Petrella (2001), pp. 61-62.

tariffs, prohibiting subsidies), affects the ownership of water as well as its regulation and its pricing.[137]

The considerable influence of the World Bank on water management by states, requires more detailed discussion. The World Bank Group consists of five institutions: the International Bank for Reconstruction and Development (IBRD), which is better known as the World Bank; the International Finance Corporation (IFC); the International Development Association (IDA); the International Centre for Settlement of Investment Disputes (ICSID); and the Multilateral Investment Guarantee Agency (MIGA).[138] Like the IMF, the World Bank was established following the Bretton Woods conference and is governed by its Articles of Agreement.[139] The World Bank's original goal was to finance the reconstruction and development of the economies of its member states. Over time, the focus of the World Bank has shifted from reconstruction to development. Among its purposes are the promotion of private foreign investment and of the long-range balanced growth of international trade (Article II(ii) and (iii) Articles of Agreement of the International Bank for Reconstruction and Development). Its principal functions are to lend funds, provide advice and technical assistance, and to serve as a catalyst to stimulate investment.

The World Bank Group has historically invested about 3 billion dollars in water-related sectors on a yearly basis and: 'Lending for water resources development and water-related services accounted for about 16 percent of all Bank lending over the past decade.'[140] The World Bank used to finance the building of large dams, but abandoned that policy due to experiences such as with the construction of three large dams in the Narmada Valley in India.[141] At present, the World Bank seems to consider re-engagement in the financing of major hydraulic infrastructure but taking into account, for example, the Report of the WCD.[142] The World Bank contributed to the development of a Comprehensive Development Framework (CDF), sup-

---

[137] See, e.g., Barlow and Clarke (2002), p. 161.

[138] The IDA forms an integral part of the World Bank. IDA and the World Bank share their staff. Moreover, both are state-oriented and aim to promote economic and social progress primarily in developing countries. The IFC has its own staff and focuses on private investors and commercial enterprises in its member states. ICSID was created in 1966 under the Convention on the Settlement of Investment Disputes between States and Nationals of Other States, which came into force on 14 October 1966, see www.worldbank.org/icsid. MIGA was established in 1988 to promote foreign direct investment into emerging economies to contribute to development and, e.g., offers political risk insurance to investors and assists developing countries attract private investment, see www.miga.org.

[139] The World Bank came into operation on 25 June 1946 and has its headquarters in Washington, D.C. The voting system provides every member with a basic vote in addition to one vote for each share it holds. Chatterjee (1992b).

[140] World Bank (2003a), p. 18.

[141] On protests resulting from problems due to the construction of the dams, such as floods and environmental implications, see Barlow and Clarke (2002), pp. 184-185.

[142] See World Bank (2003b) and www.worldbank.org/water.

porting, for example, a holistic approach, requiring that ownership must reside with communities and countries, for which approach the water sector is seen as the most naturally suited.[143] The role of the World Bank in projects involving international rivers has been of major importance.[144] The World Bank usually operates through projects, but can nevertheless provide non-project lending in special circumstances such as caused by natural disasters – for example serious flooding – that call for urgent reconstruction and restoration of an economy.[145] With regard to the projects, environmental considerations are taken into account during the appraising stage and monitored throughout the project. Such considerations can even lead to the loan being conditional upon environmental safeguards.[146]

Under the operational policies and the procedures of the World Bank on projects on international waterways, providing notice of the proposed project to the other riparians by the borrower or through the World Bank is a condition for processing the project.[147] The procedures furthermore include provisions on responses and objections from the other riparians and the possibility to seek independent expert opinion.[148]

5.5   Conclusions

It may be concluded that "ownership" of water in principle concerns user rights that are probably best regulated and controlled by democratic public bodies. State sovereignty and control over water are qualified by, for example, the rights of other states, the interests of their population and environmental obligations.

Community-public-private partnerships can under conditions provide a promising way to manage water. Within this cooperation, a central position of the community, in line with the principle of subsidiarity, could contribute to the implementation of basic access to water for all. Public regulation and control would be needed to address larger economic and ecological concerns. Any transfer of control to the private sector is to be accompanied by regulation and control safeguarding social and ecological interests.

---

[143] Pledge of The World Bank, World Water Council (2000), pp. 75-77.

[144] See Krishna (1998) on the Bank's policy for projects on international waterways.

[145] Chatterjee (1992b), p. 125.

[146] Chatterjee (1992b), p. 131.

[147] Salman and Boisson de Chazournes (1998), Annexes 2A, pp. 193-196, 2B, pp. 197-200, and 2C, p. 201, concern, respectively: The World Bank Operational Manual, Operational Policies, Projects on International Waterways, OP 7.50, October 1994; The World Bank Operational Manual, Bank Procedures, Projects on International Waterways, BP 7.50, October 1994, including under para. 1 defined types of international waterways; and The World Bank Operational Manual, Good Practices, Projects on International Waterways, GP 7.50, November 1994.

[148] On the water resources strategy of The World Bank, see www.worldbank.org/water.

An economic approach to water can assist in the efficient management of water, but is not necessarily compatible with sustainable development. When dealing with water as an economic good, regulation such as by international law is needed to ensure that the social and environmental interests are well taken into account and that the transboundary nature of water is thoroughly reflected. For international trade and investment to be compatible with sustainable development, they need to increasingly reflect fair trade and sustainable investment. States must at least be allowed to protect vital human needs and ecological sustainability by, for example, national laws and price differentiation.[149] Trade organizations such as WTO and NAFTA are not suitable organizations to arrange the management of water. As voiced by Dunoff, integration of principles of sustainable development with principles governing world trade will not occur in an institutional setting designed to further only one of these interests.[150] Moreover, institutions such as the IMF and the World Bank may be inclined to steer developing countries toward privatisation even in cases in which the desirability of privatisation is questionable.

In addition, it can be argued that the state is the custodian over water, and must take into account the common interests involved. Vital human needs, sustainability and ecosystem protection appear to be suitable candidates to receive priority in the allocation of water.

---

[149] See Section 8.4 and Annex II on pricing of water.
[150] See Dunoff (1994), furthermore arguing that trade-environment conflicts need to be dealt with by a forum sensitive to all the interests involved.

6.   Water as an ecological good

6.1   Protection of water

The ecological functions of water were identified as the other area, besides vital human needs, that needs further protection by international law if development is to be sustainable.[1] Both the ECE Convention and the Watercourses Convention address environmental concerns. Under Article 2 of the ECE Convention, parties are obliged to take all appropriate measures to prevent, control and reduce any transboundary impact, including measures to ensure the ecologically sound use of transboundary waters and environmental protection, to ensure conservation and where necessary the restoration of ecosystems, guided by the precautionary principle, the polluter-pays principle and the sustainable management of water resources. Measures to prevent, control and reduce transboundary impact are further elaborated upon in Article 3 of the ECE Convention. The application of the ECE Convention must not lead to deterioration of environmental conditions or to increased transboundary impact (Article 2.7). Guidelines for developing optimal environmental practices are included in an annex to the ECE Convention. Another example of a regional agreement arranging for the protection of water is the EU Water Framework Directive.[2]

The relevant articles of the Watercourses Convention are discussed in the following paragraphs, addressing the duty to protect water, protection of the environment and ecological integrity. Other principles of international law, categorised according to the methodology in Section 1.4, that further stress the duty of humankind to protect water as an ecological good include the principle of equitable and reasonable utilization (Section 3.3.2), intergenerational equity (Section 4.4.2), the precautionary principle (Section 7.3.2), the no-harm principle (Section 7.4.2) and the common heritage of humankind (Section 7.4.3).

The focus on the human use of water and negligence of ecological concerns has led to serious degradation of freshwater resources.[3] According to Birnie and Boyle:[4]

> Its [international water law's] principle focus, evident in the ILA's codi-
> fication of 1966 ('The Helsinki Rules'), has mainly been the rules and
> principles for allocating water supply in international watercourses be-

---

[1] See Section 2.4 on ecological functions of water. On international law and protection of international watercourses and their ecosystems, see McCaffrey (2001), Chapter 11, Tanzi and Arcari (2001), Chapter 4, Sands (2003), Chapter 10, and Birnie and Boyle (2002), Chapter 6.

[2] See Section 8.4 of this study.

[3] See also Section 2.4 of this study and Agenda 21, para. 18.35.

[4] Birnie and Boyle (2002), pp. 298-299.

tween upstream and downstream states, and only incidentally have environmental or sustainability concerns been served.

In many cases, the carrying capacity of water has been exceeded, resulting in pollutants accumulating in the aquatic food chain (bioaccumulative substances) and damage to human health.[5] Ecological degradation and disasters can be a threat to individuals as well as to states and the international community, as, for example, the Chernobyl disaster. Any kind of change in the flow and quality of watercourses, such as by dams, large diversions and pollution, can pose a possible threat to existing ecosystems, while degradation of other ecosystems can equally pose a threat to freshwater resources.[6] In some cases, degradation of freshwater resources can appear to be merely local, yet have detrimental effects in and outside the territory. Degradation of groundwater is, moreover, complicated because of its slow pace and recharge and the uncertainties as to its course and connection to surface waters.[7] The impact of pollution of groundwater, therefore, becomes unpredictable and can result in the contamination of surface water as well.

Over the last few centuries, awareness of the importance of nature for people and for its intrinsic value has increased significantly. It has become evident that taking care of people, including of future generations, entails taking care of the environment, including its life support systems. As stated by the ICJ: 'the environment is not an abstraction but represents the living space, the quality of life and the very health of human beings, including generations unborn.'[8]

The uncertain and possibly irreversible impact of water degradation emphasises the need for an integrated, transboundary and proactive approach. The ICJ in the *Gabcíkovo-Nagymaros* case stated:[9]

---

[5] See World Meteorological Organization/Stockholm Environment Institute (1997), *Comprehensive Assessment of the Freshwater Resources of the World*, UN Doc. E/CN.17/1997/9, pp. 10-11, where it is also stated that: 'Groundwaters, once contaminated, are very difficult to clean up because the rate of flow is usually slow.'

[6] See Postel and Richter (2003), on the importance and impact of river flows, and Teclaff (1994), on damage by waterworks of aquatic ecosystems and on suggested strategies for their restoration. Large diversions include bulk water transfer by pipelines as planned by Libya to relocate the waters of the Kufra basin, canals and water bags, see Barlow and Clarke (2002), pp. 133 and 137-140. See also Brown Weiss (1989), p. 237. See Section 2.3.3 of this book on dams. See Section 2.4.3 on the interaction of water with other elements of the world ecosystem.

[7] McCaffrey (2001), p. 35: 'Furthermore, aquifer systems often do not coincide with the drainage basins overlying them because of the composition and inclination of subsurface strata.' See on groundwater resources and international law, *e.g.*, Eckstein and Eckstein (2003).

[8] ICJ Advisory Opinion, *Legality of the Threat or Use of Nuclear Weapons*, ICJ Reports 1996, para. 29, as reiterated in the *Gabcíkovo-Nagymaros* case, para. 53.

[9] Gabcíkovo-Nagymaros case, para. 140.

The Court is mindful that, in the field of environmental protection, vigilance and prevention are required on account of the often irreversible character of damage to the environment and of the limitations inherent in the very mechanism of reparation of this type of damage.

Prevention is of the utmost importance and conservation is to be preferred above remedial options. As formulated by Hunter, Salzman and Zaelke: 'Preventing environmental damage is almost always less costly than allowing the damage and incurring the environmental costs and other consequences later.'[10]

6.2   Duty to protect water

A duty to protect water at the community level can also be viewed as a right to protect water against others. The environment has long been left at the mercy of social and economic human uses of water. When allocating freshwater resources, the option and value of *not* using waters is easily overlooked. Water resources entail many functions of their own and are not left useless when not exploited for human use. Even fresh water returning to the atmosphere has functions in relation to the world's ecosystems and can influence, for example, climate and biodiversity. Rivers carry with them sediments that can fertilise their banks miles downstream, but if polluted they can degrade other waters downstream, including the sea, as well.

The protection of water resources is also regarded as an essential component of sustainable development by the international community as reflected in Chapter 18 of Agenda 21 on the protection of the quality and supply of freshwater resources. This Chapter states, among other things, that legal instruments should be developed in order to protect the quality of water reserves.

In many situations, the effects of unsustainable use of water cannot easily be reversed. The required proactive approach toward protection of water has implications for policy in relation to demand and over-exploitation of water. Increased protection of water thus requires the application of environmental impact assessments before deciding upon the (abstinence from) use of water resources.

6.2.1   Increased demand

Population growth combined with the increase of demand per capita has resulted in unsustainable patterns of water consumption. As expressed by Birnie and Boyle: 'population growth and increased living standards are reflected in demand for water at rapidly increasing levels that cannot be

---

[10] Hunter, Salzman and Zaelke (2002), p. 404, on the principle of prevention.

met indefinitely'.[11] Population growth is expected to stabilise at about 9.3 billion people by the middle of the twenty-first century.[12] Until that time, population growth is expected to aggravate water problems. Population growth mainly takes place in developing countries, many of which already face tremendous problems of water allocation.

The high demand per capita, however, is mainly due to the level of consumption by people and industries in industrialised countries. According to WWDR: 'A child born in the developed world consumes thirty to fifty times the water resources of one in the developing world'.[13] As stated in the UN Freshwater Assessment: 'Water use has been growing at more than twice the rate of the population increase during this century, and already a number of regions are chronically water-short.'[14] And according to the WWDR: 'Water consumption has almost doubled in the last fifty years.'[15] It is estimated that demand will exceed the available supply of water by 56 percent by the year 2056.[16]

Reducing demand is regarded as the most promising way of conserving water.[17] The need to conserve water requires investments in reproductive health and empowerment of women and a change in lifestyle of people who now enjoy abundance.[18] In turn, this will enable freshwater resources to be developed in a sustainable fashion, which could improve the quality of life. Methods of conservation can be supplemented by other approaches such as technological development and desalination. Turning salt water into fresh water is expensive and energy-consuming, making it primarily an option for arid, oil rich countries. Other possible alternatives include the re-use of wastewater, especially by industries, and saltwater agriculture. The combination of measures to improve the productivity rather than seeking new sources is also referred to as "the soft path".[19]

---

[11] Birnie and Boyle (2002), p. 298.

[12] WWAP (2003), p. 12.

[13] WWAP (2003), p. 5.

[14] Comprehensive Assessment of the Freshwater Resources of the World, 4 February 1997, E/CN.17/1997/9. According to the assessment, two-thirds of the world population could be under water stress conditions by the year 2025.

[15] WWAP (2003), p. 5.

[16] Barlow and Clarke (2002), p. 24.

[17] See Postel (1996), p. 53, arguing that often conservation and efficiency options cost less than the development of new water resources.

[18] On population growth, see LeRoy (1995), p. 324: 'Recent research also suggests how powerfully family planning programs work in concert with improved opportunities for women, especially secondary school education for girls.'

[19] Gleick (2002) and Gleick, Burns, Chalecki, Cohen, Cushing, Mann, Reyes, Wolff and Wong (2002).

6.2.2   Replenishment rate

A condition of conserving freshwater resources is not to withdraw water beyond replenishment rates.[20] With increasing demand, water levels have declined by tens of meters in many regions where water has been pumped out beyond the natural replenishment rate for uses such as drinking water and irrigation. For example, Mexico City is drying out and has been sinking by about 50 centimetres each year.[21] In the case of fossil groundwater such as the Ogallala aquifer in the US, thousands of years old and very slowly replenished, any withdrawal is beyond such a rate of replenishment. The Ogallala aquifer is mined 14 times its replenishment rate for the irrigation of farmland.[22] Over-exploitation can, moreover, lead to degradation when it leads to salinisation.[23] As stated by Brown Weiss: 'Salt water may intrude upstream from a river mouth because fresh water no longer flows in sufficient amounts to keep sea water back.'[24]

In order to ensure that withdrawal remains within replenishment rates, it is essential to know, for example, precipitation patterns: how much water comes from the atmosphere to the surface in various areas in specific periods of time and which water systems does it replenish. It is crucial to know the flow of a watercourse, whether surface or underground, and its influence on the ecosystems it supports. In the case of the exploitation or mining of fossil freshwater resources, artificial replenishment, for example in the case of periodic drought and rain, can be an option to maintain this water at a certain level. Limiting withdrawal to replenishment rates is primarily a national policy decision but, partly depending on the size of the (sub)basin, can also involve decision-making at the community and international levels. The sustainable management of aquifers could include aquifer management organizations (AMORs), groundwater protection zones, conservation of recharge areas, granting of rights and permits, the involvement of stakeholders and the use of environmental impact assessments.

6.2.3   Environmental impact assessment

To quote Hunter, Salzman and Zaelke: 'Environmental impact assessment (EIA) is the process for assessing the impact of proposed activities, policies

---

[20] Caponera (1992), p. 247: 'In fact, if not adequately controlled, abstraction activities may cause, *inter alia*, the depletion of aquifers, the deterioration of groundwater quality, salt water intrusion in coastal areas and land subsidence.'
[21] See on the water problems of Mexico City, *e.g.*, Barlow and Clarke (2002), pp. 18-19.
[22] Barlow and Clarke (2002), p. 16. See also Glennon (2003).
[23] See Bertels, Aiking and Vellinga (1999), pp. 129-130, on intrusion of saline waters in coastal zones.
[24] Brown Weiss (1989), p. 238. At p. 236, she states: 'Ground water depletion also occurs when oil production leaves a significant portion of ground water trapped and inaccessible for future use.'

or programs to integrate environmental issues into development planning.'[25] The 1991 ECE Convention on Environmental Impact Assessment in a Transboundary Context (Espoo Convention) defines EIA as: 'a national procedure for evaluating the likely impact of a proposed activity on the environment' (Article 1(vi)).[26] According to Gillies, EIA should include an assessment of the direct and indirect effects of the project on humans, fauna, flora, soil, water, air, climate, landscape, material assets and the cultural heritage.[27] The complex interrelationships involved can be illustrated by the fact that coral reefs (made up by animals, not plants) live in symbiosis with certain fish that remove green algae from the living polyps, which would otherwise destroy the coral reefs. Over-exploitation of fish can therefore lead to the destruction of coral reefs. According to the ILA, EIA concerning transboundary effects now-a-days presents a rule of customary international law.[28]

Besides the Espoo Convention, environmental assessment is included in Article 206 UNCLOS and in such environmental treaties as the Biodiversity Convention. Article 14 of the Biodiversity Convention obliges states parties, as far as possible, to:

Introduce appropriate procedures requiring environmental impact assessment of its proposed projects that are likely to have significant adverse effects on biological diversity with a view to avoiding or minimizing such effects and, where appropriate, allow for public participation in such procedures;

According to Judge Weeramantry, in his Separate Opinion to the *Gabcíkovo-Nagymaros* case, EIA embodies the obligation to maintain watchfulness and anticipation, applying the present-day norms since it concerns not the validity of the treaty but the application.[29] This is very much in line with the Judgment in the case, which also emphasises the need for continued evaluation. Increasing knowledge of the environment and the increasing awareness of its value reaffirm the need for permanent updating of norms and monitoring and, if necessary, adaptation to protect freshwater resources.

EIA is frequently to be found in non-binding instruments, including the 1987 UNEP Goals and Principles of Environmental Impact Assessment and

---

[25] Hunter *et al.* (2002), p. 432.
[26] Convention on Environmental Impact Assessment in a Transboundary Context, Espoo, 25 February 1991, *UN Doc.* E/ECE/1250, entry into force: 10 September 1997, 1989 *UNTS*, 309, and 30 *ILM* (1991), 802. Status as of 4 October 2004: 40 parties and 30 signatories. See www.unece.org/ leginstr/cover.htm.
[27] Gillies (1999), p. 22.
[28] ILA Committee on Water Resources Law (2004), p. 31, referring to Article 3 of the ILA Supplemental Rules on Pollution and to par. 4.2 of the ILA New Delhi Declaration.
[29] Separate Opinion to the *Gabcíkovo-Nagymaros* case of Judge Weeramantry, under B (a).

the World Bank's Operational Directive on Environmental Assessment.[30] Principle 4 of the 1978 UNEP Principles on Shared Natural Resources states authoritatively: 'States should make environmental assessments before engaging in any activity with respect to a shared natural resource which may create a risk of significantly affecting the environment of another State or States sharing that resource.'[31] Principle 17 of the Rio Declaration moreover states: 'Environmental impact assessment, as a national instrument, shall be undertaken for proposed activities that are likely to have a significant adverse impact on the environment and are subject to a decision of a competent national authority.'

EIA provides an important tool for implementing environmental regulations.[32] The application of the precautionary principle will also often require EIA, such as formulated in Principle 4 of the ILA New Delhi Declaration (see 7.3.2). EIA can be applied at all levels of planning activities involving, or possibly affecting, freshwater resources, to provide insight into the requirements of conservation and protection of freshwater resources. EIA provides part of the information needed to weigh the factors of equitable and reasonable utilization in such a way as to achieve sustainable development.[33]

6.3   Protection of the environment

The Watercourses Convention includes provisions relating to environmental protection but has been criticised for not being sufficiently progressive in the field of international environmental law. McCaffrey explains: 'The fact is, they [stronger environmental provisions] *were* thought of, but were simply not acceptable to a sufficient number of delegations.'[34] The

---

[30] UNEP Goals and Principles of Environmental Impact Assessment, 17 June 1987, adopted by the Governing Council of UNEP in resolution GC14/25, see also UNGA Res. 42/184 (1987). In this instrument, EIA is defined as: 'an examination, analysis and assessment of planned activities with a view to ensuring environmentally sound and sustainable development.' World Bank's Operational Directive on Environmental Assessment, O.D. 4.01, October 1991, para. 2: 'The purpose of EA is to improve decision making and to ensure that the project options under consideration are environmentally sound and sustainable.'

[31] Draft Principles of Conduct in the Field of the Environment for Guidance of States in the Conservation and Harmonious Utilization of Natural Resources Shared by Two or More States, 19 May 1978, *ILM* 17 (1978) 1097.

[32] According to Hunter *et al.* (2002), p. 432: 'In the transboundary context, many commentators believe that the duty to conduct an EIA is probably now a requirement of customary law.' The status of the principle nevertheless is debated among states.

[33] An integrated impact assessment approach is recently taken by the EU as well, see Commission of the European Communities, *Communication from the Commission on Impact Assessment*, Brussels, 5.6.2002, COM(2002) 276 final.

[34] McCaffrey (1998), p. 27.

several environmental elements included in the Watercourses Convention can be viewed as minimum standards for an equitable regime.[35]

The ICJ in its 1997 *Gabcíkovo-Nagymaros* judgment took ecological concerns into account, although in this case it ruled that Hungary could not justify the breach of its treaty with Slovakia on the ground of ecological necessity.[36] The ICJ in its Advisory Opinion on Nuclear Weapons affirmed that, to a certain extent, the environment is protected also in times of war: 'Respect for the environment is one of the elements that go to assessing whether an action is in conformity with the principles of necessity and proportionality.'[37] Concerning the Additional Protocol I to the Geneva Conventions[38] the ICJ stated:[39]

> Taken together, these provisions embody a general obligation to pro-
> tect the natural environment against widespread, long-term and severe
> environmental damage; the prohibition of methods and means of war-
> fare which are intended, or may be expected, to cause such damage;
> and the prohibition of attacks against the natural environment by way

---

[35] Tanzi and Arcari (2001), pp. 225-232, state at p. 231, that the environmental provisions of the Watercourses Convention codify the 'lowest common denominator'.

[36] The Court did not establish a short-term risk of an ecological disaster, caused by the Gabcíkovo-Nagymaros project. The ICJ also took into account the contribution of Hungary to the emergency situation by refusing to discuss adaptations of the project with Slovakia after 1991. See Lammers (1998) for a thorough discussion of the *Gabcík-ovo-Nagymaros* case from the perspective of international watercourses and environmental protection, including the state of (ecological) necessity. See also Sands (1998) who concludes that the Court has gone some considerable way towards developing the law in relation to watercourses and the need to protect the environment.

[37] ICJ Advisory Opinion, *Legality of the Threat or Use of Nuclear Weapons*, ICJ Reports 1996, 226, para. 30. In para. 32, the ICJ refers to its Order in the *Request for an Examination of the Situation in Accordance with Paragraph 63 of the Court's Judgment of 20 December 1974 in the* Nuclear Tests (New Zealand v. France) Case, ICJ Reports 1995 p. 306, para. 64, in which it stated that its conclusion was 'without prejudice to the obligations of States to respect and protect the natural environment'. The ICJ finds in para. 33 that the existing international law relating to the protection and safeguarding of the environment 'indicates important environmental factors that are properly to be taken into account in the context of the implementation of the principles and rules of the law applicable in armed conflict.'

[38] Articles 35, paragraph 3, and 55 of the 1977 Additional Protocol I to the 1949 Geneva Conventions.

[39] ICJ Advisory Opinion, *Legality of the Threat or Use of Nuclear Weapons*, ICJ Reports 1996, 226, para. 31. Reference was furthermore made, in para. 30, to Principle 24 of the Rio Declaration: 'Warfare is inherently destructive of sustainable development. States shall therefore respect international law providing protection for the environment in times of armed conflict and cooperate in its further development, as necessary.' And in para. 31 to UNGA Resolution 47/37 of 25 November 1992 on the Protection of the Environment in Times of Armed Conflict which states that 'destruction of the environment, not justified by military necessity and carried out wantonly, is clearly contrary to existing international law'.

of reprisals. These are powerful constraints for all the States having subscribed to these provisions.

The provisions of Part IV – on the protection, preservation and management of international watercourses – of the Watercourses Convention are of specific relevance. Its articles will be separately discussed in this Chapter: Article 20 on the protection of ecosystems; Article 21 on pollution; Article 22 on alien or new species; and Article 23 on the marine environment.[40]

6.3.1   Protection of ecosystems

The ILC in its comment to Article 20 of the 1994 Draft Articles of the Watercourses Convention states that "ecosystem" generally 'refers to an ecological unit consisting of living and non-living components that are interdependent and function as a community.'[41] Article 2 of the UN Convention on Biological Diversity furthermore states: 'Ecosystem means a dynamic complex of plant, animal, and micro-organism communities and their non-living environment interacting as a functional unit.' It can be argued that states are obliged under international law to protect freshwater resources and their ecosystems.[42]

The obligation to protect ecosystems is part of the Watercourses Convention.[43] Article 20 of the Watercourses Convention obliges watercourse states to protect and preserve the ecosystems of international watercourses: 'Watercourse States shall, individually and, where appropriate, jointly, protect and preserve the ecosystems of international watercourses.' The duty to protect is an application of the precautionary approach.[44] The obligation to preserve mainly relates to freshwater ecosystems in their original condition.[45] Article 20 is not limited to transboundary issues in the sense in

---

[40] These articles are largely based upon Articles 192, 194, 196 and 207 of UNCLOS, see Tanzi and Arcari (2001), pp. 232-234.

[41] ILC 1994 Draft Articles on the Law of the Non-navigational Uses of International Watercourses, p. 99, see www.un.org/law/ilc/texts/94nonnav.pdf. See also Report of the International Law Commission on the Work of its Forty-Sixth Session, *UN Doc.* A/49/10 (1994), pp. 280-281.

[42] McCaffrey (2001), p. 396: 'While this obligation [to protect the ecosystems of international watercourses] may be described as 'new' or 'emerging', its basic elements are already part of general international law.'

[43] On the process that led to the inclusion of the term ecosystem in the Watercourses Convention, see Tanzi and Arcari (2001), pp. 238-242.

[44] Report of the International Law Commission on the Work of its Forty-Sixth Session, *UN Doc.* A/49/10 (1994), p. 119.

[45] Report of the International Law Commission on the Work of its Forty-Sixth Session, *UN Doc.* A/49/10 (1994), p. 282:

'while similar to that of protection, [preserve] applies in particular to freshwater ecosystems that are in pristine or unspoiled condition. It requires that these ecosys-

which equitable and reasonable utilization and the no-harm principle are normally limited.

An ecosystem approach is also reflected in Article 22 on alien or new species and Article 23 on the marine environment of the Watercourses Convention.

6.3.2   Combating pollution of water

Pollution, both by point and diffuse sources, has become an increasingly acute problem. For example, in the area of the Great Lakes, located on the border between the US and Canada, an estimated 50 to 100 million tons of hazardous waste is generated yearly.[46] Pollution, such as caused by agricultural, industrial and domestic sewage effluents, is not solely a matter of water quality but also of water quantity and its specific use.[47] According to Brown Weiss: 'Persistent toxic contamination of streams and lakes kills plants, fish, and other forms of animal life and makes impossible some water uses, such as drinking and swimming.'[48] Bioaccumulation causes higher levels of toxins at each level of the food chain. Moreover, wastewater treatment is not always available and, in cases where it is, it is not always adequate to safeguard human health or that of aquatic organisms.[49]

International law has dealt extensively with pollution, *e.g.* through the 1999 Convention on the Protection of the Rhine and the 1989 Convention on the Control of Transboundary Movements of Hazardous Wastes and their Disposal.[50] Pollution *per se* appears not to be prohibited under international law, but it is limited by it.[51] In Article 21 on the prevention, reduction and control of pollution, the Watercourses Convention states:

---

tems be protected in such a way as to maintain them as much as possible in their natural state.'

[46] Barlow and Clarke (2002), p. 35.

[47] As stated by Bertels *et al.* (1999), pp. 132-139: 'As water circulates through the biosphere, it is susceptible to pollution from many sources.' Water quantity influences the concentration of pollutants. Caponera (1992), p. 153: 'A water body may be polluted for one purpose such as drinking, but not for another (industrial, irrigation, etc.).'

[48] Brown Weiss (1989), pp. 232-233.

[49] In developing countries, about 90 per cent of waste water is discharged without treatment. On concerns about estrogenic chemicals in the water not sufficiently removed by waste water treatment plants that might cause reproductive problems for aqautic organisms, see Vethaak, Rijs, Schrap, Ruiter, Gerritsen and Lahr (2002).

[50] On the International Commission for the Protection of the Rhine guarding the convention, see Birnie and Boyle (2002), pp. 324-326. Convention on the Control of Transboundary Movements of Hazardous Wastes and their Disposal, Basel, 22 March 1989, entry into force: 5 May 1992, 28 *ILM* (1989), 657. Status as of 4 October 2004: 163 parties and 53 signatories. In general on international law and pollution of freshwater resources, see Nollkaemper (1993) and Lammers (1984).

[51] See Birnie and Boyle (2002), p. 330. According to the commentary to Article 27 of the Berlin Rules: 'the obligation to control pollution in order to produce the least net

1. For the purpose of this article, 'pollution of an international watercourse' means any detrimental alteration in the composition or quality of the waters of an international watercourse which results directly or indirectly from human conduct.

2. Watercourse States shall, individually and, where appropriate, jointly, prevent, reduce and control the pollution of an international watercourse that may cause significant harm to other watercourse States or to their environment, including harm to human health or safety, to the use of the waters for any beneficial purpose or to the living resources of the watercourse. Watercourse States shall take steps to harmonize their policies in this connection.

3. Watercourses States shall, at the request of any of them, consult with a view to arriving at mutually agreeable measures and methods to prevent, reduce and control pollution of an international watercourse, such as:

(a)   Setting joint water quality objectives and criteria;

(b)   Establishing techniques and practices to address pollution from point and non-point sources;

(c)   Establishing lists of substances the introduction of which into the waters of an international watercourse is to be prohibited, limited, investigated or monitored.

Article 21.1 thus excludes natural pollution, but it refers to human conduct such that not only introduction of pollutants are included but potentially any human conduct causing pollution. An increase in the threat posed by pollutants requires a higher level of due diligence, possibly even resulting in a strict obligation.[52] The formulation 'may cause' in Article 21.2 indicates that early action should be taken. In its commentary, the ILC specified Article 21 as 'a specific application of the general principles contained in Articles 5 and 7.'[53] It would appear that significant pollution caused by a specific use, which causes harm to other watercourse states, constitutes an inequitable and unreasonable use.[54]

Secondary sources of international law dealing with water pollution include the work of the ILA in the 1966 Helsinki Rules and the 1982 Montreal Rules.[55] According to the ILA definition in Article 9 of the Helsinki Rules, water pollution is: 'Any detrimental change resulting from human

---

environmental harm is part of the customary international law of the environment.', ILA Committee on Water Resources Law (2004), p. 30.

[52] McCaffrey (2001), pp. 386-387.

[53] Report of the International Law Commission on the Work of its Forty-Sixth Session, *UN Doc.* A/49/10 (1994), p. 291.

[54] See McCaffrey (2001), pp. 385-386. On the no-harm principle, see Section 7.4.2 of this study.

[55] ILA Montreal Articles on Water Pollution in an International Drainage Basin, 1982 ILA Report of the Sixtieth Conference held at Montreal.

conduct in the natural composition, content or quality of the waters of an international drainage basin.' National law has also extensively dealt with pollution.[56]

The interrelationship of freshwater resources with, for example, the sea, and between surface water and groundwater calls for an integrated approach in tackling pollution. Moreover, in cases where only certain forms of pollution are controlled, there can be a shift in pollution, *e.g.* from water to air, which does not solve the problem but merely relocates it. An example of an integrated approach toward pollution is provided by the EU Directive on Integrated Pollution Prevention and Control.[57]

6.3.3   Alien or new species

The introduction of alien or new species is often referred to as a form of pollution. These species have been introduced by, for example, ballast water from other regions carried by ships.[58] Outside their natural biotope, species may have no natural enemies and may dominate or at least strongly influence their new environment. For example, in the Great Lakes basin the introduction of the zebra mussel by a ship's ballast in 1988 has almost extinguished the plankton needed by native fish and mussels.[59] "Species" were defined by the ILC as flora as well as fauna, "alien" refers to non-native and "new" to genetically altered or created through biological engineering.[60]

Article 22 of the Watercourses Convention obliges watercourse states to take measures to prevent the introduction of alien or new species that might cause significant harm:

> Watercourse States shall take all measures necessary to prevent the introduction of species, alien or new, into an international watercourse which may have effects detrimental to the ecosystem of the water-course resulting in significant harm to other watercourse States.

According to Tanzi and Arcari:[61]

> However, the reference to the no harm rule in Article 22 appears to be inconsistent with the ecosystemic approach which emerges from the obligation of protection and preservation of Article 20, which has not been made dependent on harm being caused to riparian States.

---

[56] See Caponera (1992), pp. 256-258 on measures taken by states to combat pollution.
[57] On the Directive, see Gillies (1999), pp. 23 and 25.
[58] See on ballast water and biosecurity, *e.g.*, McGee (2002).
[59] See Barlow and Clarke (2002), p. 43.
[60] See Tanzi and Arcari (2001), p. 271, citing Report of the International Law Commission on the Work of its Forty-Sixth Session, *UN Doc.* A/49/10 (1994), p. 297.
[61] Tanzi and Arcari (2001), pp. 273-274.

6.4   Ecological integrity

States can be argued to bear a global environmental responsibility, a responsibility to safeguard the ecological integrity of the earth.[62] Article 3.6 of the Berlin Rules defines ecological integrity as 'the natural condition of waters and other resources sufficient to assure the biological, chemical, and physical integrity of the aquatic environment.' Article 22 on ecological integrity states: 'States shall take all appropriate measures to protect the ecological integrity necessary to sustain ecosystems dependent on particular waters.' According to the commentary to this Article this obligation 'has only recently been recognized in international and national legal systems, but has rapidly gained general acceptance.' As formulated by McCaffrey:[63]

> While problems of water pollution have perhaps received more attention in the literature, it seems probable that the protection of watercourse ecosystems is of wider significance, in terms of geography, meeting basic human needs, and sustainable development.

Elements of ecological integrity can be found, for example, in the ICJ *Advisory Opinion on Nuclear Weapons* and the Separate Opinion of Judge Weeramantry to the *Gabcíkovo Nagymaros* case.[64] Agenda 21 includes the objective of: 'Maintenance of ecosystem integrity, according to a management principle of preserving aquatic ecosystems, including living resources, and of effectively protecting them from any form of degradation on a drainage basin basis.'

The geographical characteristics of water resources imply that the catchment basin approach is most suitable to arrange for the ecologically sound management of water resources. In addition, the interdependency of water resources involves the oceans as well. The various threats to ecosystems and humans alike, as well as their interrelationship, point to the larger mutual interdependency underlying the concern for ecological integrity.

6.4.1   Catchment basin approach

The catchment basin approach was identified and supported by the findings of the foregoing chapters as the appropriate geographical level to manage water resources.[65] As stated by McCaffrey: 'Gravity pulls water

---

[62] Birnie and Boyle (2002), pp. 97-104, provide an elaboration on the environment as a common concern. See on ecological integrity, Karr (1993) and Woodley, Kay and Francis (1993).

[63] McCaffrey (2001), p. 396.

[64] See Section 6.3 of this study.

[65] Partly depending on geographical and political practicalities, in specific cases a sub-basin approach can be more appropriate. According to the commentary to Article 2 of the Berlin Rules: 'Management planning normally wil be based upon drainage basins, although there might be particular reasons for developing plans on some other basis,

ever in a downhill direction, from the mountains to the sea, through the catchment basins into which it falls. These basins themselves form systems, whose component parts are interrelated.'[66] A catchment basin does not necessarily involve (much) surface water. For example, precipitation can fall into zones such as deserts where all of it either seeps into the soil, becoming groundwater, or evaporates. A trend toward the basin approach can be identified in international instruments but it is not yet an established principle within international law.[67] In practice, a catchment basin or area is very similar to modern use of "river basin".

Also the definitions of "watercourses" and "international watercourses" employed in international law instruments such as the ECE Convention and Watercourses Convention result in approaches very similar to that of the basin and international basin respectively. Article 2 of the ECE Convention moreover refers to catchment areas. A basin approach is also taken by various regional water treaties, including the 1969 Plata River Basin, the 1978 Great Lakes Water Quality Agreement, the 1987 Zambezi Action Plan and the 1995 Mekong Agreement.[68] Remarkably many of these treaties concern regions in the developing world. According to Birnie and Boyle:[69]

Environmental protection arrangements in Europe and North America are incomplete, apply only to certain rivers, and have only slowly been implemented. African watercourse treaties are sophisticated in content, but of little practical importance due to their limited implementation.

Article 2 of the 1966 ILA Helsinki Rules defines the international catchment basin as follows: 'An international drainage basin is a geographical area extending over two or more States determined by the watershed limits of the system of waters, including surface and underground waters, flowing into a common terminus.'[70] Many if not most definitions of basins

---

such as parts of a basin or combining all or parts of several basins.', ILA Committee on Water Resources Law (2004), p. 9. See on the river basin, Teclaff (1996) and (1967). The feasability of a basin approach remains subject of debate, see Wester and Warner (2002).

[66] McCaffrey (2001), p. 52.

[67] According to Teclaff (1996), p. 390: 'The river basin concept is, in itself, a precautionary principle, capable of sustaining us in the 21st Century, soundly established in theory and gaining in practice.'

[68] Parties to the 1969 La Plata Basin Treaty are Argentina, Bolivia, Brazil, Paraguay and Uruguay. 1978 Great Lakes Water Quality Agreement is an agreement between Canada and the United States. 1987 Action Plan for the Environmentally Sound Management of the Common Zambezi River System, agreement between Botswana, Mozambique, Tanzania, Zambia and Zimbabwe, United Nations 1987. See on the Mekong Agreement Section 3.2.3.

[69] Birnie and Boyle (2002), pp. 330-331.

[70] Similar to Article 2.1 ILA Campione Consolidation. The 1966 Helsinki Rules was the first instrument of international law to refer to the common terminus.

in instruments of international law, including the Watercourses Convention, also refer to a common terminus. However, the rapid growth of understanding of the hydrological cycle has made the term "common terminus" outdated.[71] Since actual confined water appears to be a rare exception at the least, about all water could nevertheless be included if a broad interpretation is employed. The possible application of altered visions or new principles on existing regimes is, moreover, accepted by the ICJ in the *Gabcíkovo-Nagymaros* case. Both the terms "common terminus" and "confined" are probably best avoided in the negotiation of new instruments in order to enable a comprehensive and integrated approach.

6.4.2 Marine environment

Freshwater resources and the marine environment are ultimately connected through the global hydrological cycle. In coastal zones the chain of pollution passes from rivers to oceans. Nor are species always restricted to either freshwater or marine environments. Migratory species move from one to the other.

Since fresh water resources interact with the sea, their management can be influenced by the obligations under the law of the sea. Freedom of the high seas is part of customary international law and a cornerstone of international law, laid down in UNCLOS. The high seas are open to all states, whether coastal or land-locked (Article 87 UNCLOS), and are to be reserved for peaceful purposes (Article 88 UNCLOS).[72] However, the Area – consisting of the seabed and ocean floor and subsoil thereof, beyond the limits of national jurisdiction – and its resources are stated to be the common heritage of mankind (Article 136 UNCLOS). In the Area activities are to be carried out for the benefit of humankind (Article 140 UNCLOS). Part XII of UNCLOS obliges parties to protect and preserve the marine environment, requiring global and regional cooperation. States are, moreover, under a duty to take measures for and cooperate in the conservation of the living resources of the high seas.[73] In addition, protection is provided by Article 23 of the Watercourses Convention, which states:

> Watercourse States shall, individually and, where appropriate, in cooperation with other States, take all measures with respect to an international watercourse that are necessary to protect and preserve the ma-

---

[71] McCaffrey (2001), p. 39: 'the idea of a 'common terminus' (...) is difficult to reconcile with common phenomena such as the extensive deltas of large rivers, and discharge into the sea of water from a watercourse system through aquifers'; at p. 40, he furthermore states that the 'common terminus' is recognized to be hydrologically incorrect.

[72] See also Part X UNCLOS, on the right of access of land-locked states to and from the sea and freedom of transit.

[73] Articles 117-120 UNCLOS.

rine environment, including estuaries, taking into account generally accepted international rules and standards.

Articles 23 poses obligations of due diligence.[74] This separate provision on the relationship between freshwater courses and the marine environment, stresses the importance of the latter and encourages an integrated approach.[75]

6.4.3   Interdependency

Awareness of the larger interdependency between humankind and its environment is growing but its reflection in international law remains restrictive. As formulated by Birnie and Boyle, the present principle of equitable and reasonable utilization:[76]

> affords an insufficient basis for measures of more comprehensive environmental protection. Nor does it ensure the integration of ecological, developmental, and intergenerational considerations which is central to sustainable development as the overriding objective of contemporary water resources policy.

As concluded in Chapter 3, equitable and reasonable utilization and its outcomes need adaptation conforming to the goal of sustainable development. A serious drought in one country can lead to increasing numbers of refugees and immigrants in other states; while the exploitation of aquifers can cause water levels in a neighbouring country to drop. In the 1982 World Charter for Nature, the UNGA recognised terrestrial and marine ecosystems to be "life-support systems".[77] "Life" refers of course not only to human life but includes all living organisms. For example, water systems provide habitats for wildlife and the migration of species are very much linked to water.[78] According to McCaffrey:[79]

> As understanding of the interactions between various species and natural systems increases, it seems inevitable that states in their practice will recognize an expansion of both the notion of the watercourse ecosystem and the legal protection thereof.

---

[74] Report of the International Law Commission on the Work of its Forty-Sixth Session, *UN Doc*. A/49/10 (1994), p. 122, para. 6.

[75] See on Article 23, *e.g.*, Tanzi and Arcari (2001), pp. 277-278.

[76] Birnie and Boyle (2002), p. 330.

[77] 1982 World Charter for Nature, *UN Doc*. A/RES/37/7 (XXXVII), 22 *ILM* 455 (1983).

[78] See Section 2.4 of this study and Birnie and Boyle (2002), p. 330.

[79] McCaffrey (2001), p. 393.

According to Brunnée and Toope, no freshwater regime is likely to promote environmental security in the long-term without ecosystem-oriented principles such as sustainable development, intergenerational equity, precaution, common concern, and the drainage basin focus.[80]

6.5 Conclusions

Water as an ecological good requires better protection by international law, including enhancement of the catchment area approach. The catchment basin approach increasingly receives support but cannot be said to be firmly established as a principle of international law. At the community level, a duty to protect water can be identified. At the national and international level, the environmental provisions in the Watercourses Convention can be viewed as minimum standards of international law. The provisions on the protection, preservation and management of international watercourses in Part IV of the Watercourses Convention deal with the protection of ecosystems, pollution, the introduction of alien or new species, and protection of the marine environment. In addition, the no-harm principle offers protection of the environment but only refers to significant transboundary harm beyond the jurisdiction of the state of origin. States may increasingly be expected to bear a responsibility to safeguard the ecological integrity of the earth.

The process of equitable and reasonable utilization as formulated in the Watercourses Convention is inconclusive on the protection offered to ecosystems and water. The separate articles in the Watercourses Convention on ecosystem protection emphasise its importance but probably are interpreted merely as factors in the process of equitable and reasonable utilization. Moreover, the principle of equitable and reasonable utilization as codified in the Watercourses Convention limits itself to watercourse states. Amendments of the principle of equitable and reasonable utilization are required in its understanding of "use" and of "international interest" to include abstinence from use and to expand to non-riparian states as well as non-state actors and to freshwater resources internationalised by their relationships within the global water system and with other ecosystems.

Preventive protection of water is to be preferred considering that it is more cost-effective than trying to combat the degradation of water, which may be difficult to reverse. Such preventive action calls for environmental impact assessments. Preventive protection of water resources can in certain cases imply abstinence of water use. This depends on, for example, the carrying capacity of the environment. The value of not using water is to be reflected when considering water an economic good in order to stimulate sustainable use of water.[81]

---

[80] Brunnée and Toope (1997).
[81] See also Section 7.4 of this study.

PART III. Balancing the Pillars of Sustainable Development

7. Water as a social, economic and ecological good

7.1 Balance of interests

The previous chapters have classified the principles in terms of the pillars of sustainable development. A common element through the research is that sustainable development is best served by situations that allow for a balance of interests. During processes of decision-making – negotiating, planning, re-evaluating or implementing instruments of law – compromises are frequently involved. Nevertheless, where trade-offs present a bias toward either social, economic, or ecological interests, they may obstruct the achievement of sustainable development and in the long-term cannot lead to intra- or intergenerational equity. Identifying the common interest between the pillars of sustainable development enables finding compromises that eliminate a bias against one or more of the pillars.

This Chapter makes three sets of combinations of the pillars of sustainable development: social and economic, social and ecological, and economic and ecological.[1] In relation to each of these combinations, it identifies the conflict of interests and the common interest. The conflict of interests exposes the challenges inherent in the search for a balance between the pillars of sustainable development. The common interest shows the opportunities for combining the various dimensions of the pillars of sustainable development in terms of principles of international law. The social and economic interests relating to water primarily address issues of *development through water*. Both social and ecological interests relate to *life support by water*. Economic and ecological interests find their common denominator in the *sustainable use of water*. And, finally, the combination of all social, economic and ecological interests in water requires *guardianship over water* (see Chapter 8).

The table below presents a summary of the conflicts of interests and the common interests in the three combinations of the pillars of sustainable development discussed in this Chapter. The conflicts of interests and the common interests are analysed in relation to the community, national and international level.

---

[1] The term "combination" is used here as a unification of component elements that remain individually distinct while "integration" combines the elements but also takes a step further in forming a whole. In Chapter 8, all pillars and principles are combined.

	Social – Economic	Social – Ecological	Economic – Ecological
Conflict of Interests ▪Community level: ▪National level: ▪International level:	*solidarity vs individuality* •local *vs.* global •national *vs.* international •intra-generational *vs.* intergenerational	*utilisation vs. conservation* •demand *vs.* availability •sovereignty *vs.* integrity •population *vs.* ecosystems	*freedom vs. responsibility* •rights *vs.* duties •independence *vs.* interrelationship •*aquae liberum vs. aquae clausum*
Common Interest ▪Community level: ▪National level: ▪International level:	*development through water* •right to development •right of self-determination •common but differentiated responsibilities	*life support by water* •right to a healthy environment •precautionary principle •eco-justice	*sustainable use of water* •polluter and user pays principle •no-harm principle •common heritage or concern of humankind

7.2   Social-economic: development through water

Development through water is identified as the over-arching concept needed to resolve the dilemma between the social and the economic pillars of sustainable development: *solidarity vs. individuality*.[2]

At the community level, this dilemma can be expressed by local *vs.* global, that is to say community *vs.* national or international interests. The significance of the community level has been underlined throughout the preceding research. The identity and culture of a community are valuable and can be threatened by unbridled globalisation, namely where globalisation imposes uniform rules that communities do not recognise as their own. On the one hand, as stated by Hey: 'Most of us participate in the process of globalization – at least as consumers – and all of us may experience its effects'.[3] On the other hand, individual autonomy and the accompanying freedom to decide matters affecting ones own life in many ways defines human beings. In the application of human rights, such freedoms are to be respected but within the limits of other people's freedoms and,

---

[2] Individuality is categorised as 'economic' considering the worldwide liberal tendency at present, in which the individual is placed at the centre.

[3] Hey (2003), p. 2, quoting her earlier publication 'Globalization and International Organizations', introduction to the 'Recurring Themes' column, 2 *International Law FORUM du Droit International* 2000, p. 220, which issue deals with globalisation as does issue 4 *International Law FORUM du Droit International* 2002.

moreover, within the boundaries of collective interest. The global community has an interest in the global hydrological cycle, including the sea. The necessary conciliation is indicated in the expression 'act locally, think globally.' While applying the subsidiarity rule and leaving room for local differences, the global context has to be taken into account; while globalisation has to integrate sustainable development.[4]

At the national level, the apparent conflict of interests can be expressed by the terms national *vs.* international. Although sovereignty may have changed in its content and application, the state remains an important player in the international field. To the extent that international organizations are composed of states, a certain degree of legitimacy and democracy can be guaranteed through state participation. Nevertheless, the international level tends not to reflect such democracy and as a result can lead to the estrangement of people as the powers of such institutions grow. On the other hand, concepts such as that of implied-powers enable international organizations to supersede national interests and to concentrate more on common interests, which naturally includes interests in the hydrological cycle. In providing a balance between national and international interests in freshwater resources, parallels can be drawn with the law of the sea as laid down in the 1982 UN Convention on the Law of the Sea (UNCLOS). States could manage the water within their territory, when appropriate, in cooperation with others, and the hydrologic system as a whole could be guarded at the international level. An international coordinating institution, dealing with the global commons, might set the goals and enable exchange of information and expertise, as is needed for fact-finding, comparable to the Authority guarding the Area, defined in Article 1 UNCLOS as the sea-bed and ocean floor and subsoil thereof, beyond the limits of national jurisdiction. As stated by McCaffrey:[5]

> The international community has devised a system for sharing the resources of the sea with developing and geographically disadvantaged states. It would seem equally important that it begin the elaboration of a system for the sharing of the world's freshwater resources equitably among all states, especially those that are *hydrologically* disadvantaged.

---

[4] In its call for sustainable development, the World Commission on the Social Dimension of Globalization (2004), p. ix, states: 'The quest for a fair globalization must be underpinned by the interdependent and mutually reinforcing pillars of economic development, social development and environmental protection at the local, national, regional and global levels.'

[5] McCaffrey (2001), pp. 172-173. A further definition of 'hydrologically disadvantaged states' seems required, possibly including desert regions, heavily polluted areas and arid but highly populated regions. A 'sharing of water resources' could include trade in products using a lot of water (virtual water trade), sharing of the benefits and transfer of technology.

At the international level, the conflict of interests can be expressed as intra-generational *vs.* intergenerational. On the one hand, as long as even basic human needs of a large part of the present generation are not fulfilled, the preservation of the environment for future generations is too much to ask of those people, particularly where this request comes from industrialised countries that have already developed themselves at great costs to the environment and do not seem to be willing to seriously reduce their own consumption. Moreover, without the present generation there can be no future generations. On the other hand, the survival and dignity of humankind require increased consideration for the interests of future generations. In applying sustainable development, the developmental side has to be given the kind of attention that is translated into action. The interdependence among people, and their dependence on the environment, as well as principles of justice, are reflected in such principles as common but differentiated responsibilities. Furthermore, especially in areas of abundance, a change in mentality is required, reflecting the relation between self-interest and global interest in time as well as an appreciation of "sufficient" instead of "more" that could reduce scarcity in many ways and thereby contribute to a possible reconciliation of present and future generational needs.

Development through water can be translated into the following international law principles in response to the dilemmas at respectively the community, national and international level: the right to development; the right of self-determination; and common but differentiated responsibilities.

7.2.1   The right to development

The right to development is often viewed as a human right.[6] Although the human rights treaties do not contain an explicit reference to a right to development, this right can be found in many other instruments. Article 1(1) of the 1986 UNGA Declaration on the Right to Development (DRD) defines the right to development as:[7]

---

[6] In the report by Mr. Arjun Sengupta, *Fourth report of the independent expert on the right to development*, UN Economic and Social Council, Commission on Human Rights, Working Group on the Right to Development, E/CN.4/2002/WG.18/2 of 20 December 2001, p. 3, para. 2, reference is made to the definition of the right to development as 'the right to a particular process of development in which all human rights and fundamental freedoms can be fully realized'. The first to conceptualise the right to development was Kéba M'Baye in his Strassbourg lecture, see M'Baye (1972), pp. 503-504. On the right to development, see Cassese (2001), pp.401-402, Chowdhury, Denters and De Waart (1992), De Waart, Peters and Denters (1988), Part 7, and Crawford (1988).

[7] See Section 3.2.2, note 28. See on the *travaux préparatoires* of the DRD, Kenig-Witkowska (1988).

an inalienable human right by virtue of which every human person and all peoples are entitled to participate in, contribute to and enjoy economic, social, cultural and political development, in which all human rights and fundamental freedoms can be fully realized.

Therefore, access to water not only constitutes a condition to the fulfilment of the right to development, the right to development in its turn can be said to enable the realisation of all human rights. The right to development has been dealt with as a right of states and/or as a human right in various documents, in both situations requiring access to water.[8]

According to Nayak, the right to development is based on the metamorphosis of international law from a law of co-existence to a law of co-operation and the emergence of mankind as a proper subject of international law.[9] Nevertheless, within international law the right to development, a trend toward cooperation, and participation of non-state actors are developing but cannot be said to be that well-established.

Article 8 of the DRD provides that: 'States should undertake, at the national level, all necessary measures for the realization of the right to development and shall ensure, *inter alia*, equality of opportunity for all in their access to basic resources.' In its interpretation of Article 8 of the DRD, the persistent conditions of underdevelopment of people, without adequate access to essentials such as food, water, clothing, housing and medicine, are referred to by the UN as a mass violation of human rights.[10] The right to development can also be found in Article 22 of the African Charter on Human and Peoples' Rights.[11] The 1986 ILA Seoul Declaration presents another example, describing the right to development as follows: 'The right to development is a principle of public international law in general and of human rights law in particular, and is based on the right of self-

---

[8] According to Bulajić (1988), p. 359, the right to development 'may be seen, both for individuals and for States, as a right of access to the means necessary for realizing human rights... and as a corollary of the right to self-determination.' See on the relation between the right to development as a State right, within the context of the New International Economic Order, and as a human right, De Waart (1988).

[9] Nayak (1992), p. 146, where he furthermore regards the duty of states to cooperate for the advancement of world peace, progress, prosperity and solidarity the fundamental source of the right to development.

[10] *The United Nations and Human Rights 1945-1995*, United Nations Blue Book Series, Vol. II, Department of Public Information, United Nations Publications: New York, as cited in Gleick (2000), p. 9.

[11] Article 22 of the African Charter on Human and Peoples' Rights states: '1. All peoples shall have the right to their economic, social and cultural development with due regard to their freedom and identity and in the equal enjoyment of the common heritage of mankind. 2. States shall have the duty, individually or collectively, to ensure the exercise of the right to development.' See also note 17 of Section 3.2.1 of this study.

determination of peoples'.[12] It is noteworthy, though, that this was not actually repeated by the ILA in its New Delhi Declaration.

The two main conferences on sustainable development, UNCED and the WSSD, also refer to the right to development. Principle 3 of the Rio Declaration states: 'The right to development must be fulfilled so as to equitably meet developmental and environmental needs of present and future generations.'[13] The 1993 Vienna Declaration on Human Rights reaffirms the right to development.[14] The introduction of the Plan of Implementation of the WSSD underlines this: 'Peace, security, stability and respect for human rights and fundamental freedoms, including the right to development, as well as respect for cultural diversity, are essential for achieving sustainable development and ensuring that sustainable development benefits all.'[15]

Despite the WSSD reaffirmation in the Plan and strong support in literature, there has also been criticism of the right to development arguing, for example, that it represents a goal more than a right.[16]

7.2.2   The right of self-determination

Self-determination of peoples is laid down as a principle in Article 1.2 of the UN Charter and as a right in the identical Articles 1 of the 1966 ICESCR and ICCPR, which state:

1.   All peoples have the right of self-determination. By virtue of that right they freely determine their political status and freely pursue their economic, social and cultural development.

2.   All peoples may, for their own ends, freely dispose of their natural wealth and resources without prejudice to any obligations arising out of international economic co-operation, based upon the principle of mutual benefit, and international law. In no case may a people be deprived of its own means of subsistence.

3.   The States Parties to the present Covenant, including those having responsibility for the administration of Non-Self-Governing and Trust Territories, shall promote the realization of the right of self-

---

[12] Principle 6, para. 1, 1986 ILA Declaration on the Progressive Development of Principles of Public International Law relating to a New International Economic Order, ILA Report of the sixty-second Conference, Seoul 1986, pp. 2-12.

[13] According to Boyle and Freestone (1999), p. 11: 'Principle 3 of the Rio Declaration is the first occasion on which the international community has fully endorsed the previously controversial concept of a 'right to development'.' They nevertheless continue, at p. 12, to say that the legal status of the right to development remains doubtful.

[14] 1993 World Conference on Human Rights, Vienna Declaration and Programme of Action, *UN Doc.* A/CONF.157/23 (1993), para. 10 and 11.

[15] WSSD Plan of Implementation, para. 5.

[16] Such criticism came, *e.g.*, from the USA, who made a reservation with regard to Principle 3 of the Rio Declaration.

determination, and shall respect that right, in conformity with the provisions of the Charter of the United Nations.

The right of self-determination of peoples thus grants peoples the right to dispose freely of their natural resources and, most importantly, it states that a people may in no case be deprived of its own means of subsistence, which includes access to water.

That resources such as water are to be used by a state in the interest of the people of that state is furthermore expressed in, for example, the first paragraph of the 1962 Declaration on Permanent Sovereignty over Natural Resources, Article 7 of the 1974 Charter of Economic Rights and Duties of States and in Article 21 of the African Charter on Human and Peoples' Rights.[17] The ICJ in its judgment in the 1995 *East Timor* case, and in its 2004 Advisory Opinion on the construction of a wall by Israel on Palestinian territory, qualified the right of peoples of self-determination as *erga omnes*.[18]

7.2.3   Common but differentiated responsibilities

The fact that there is a common interest in the hydrological cycle is not to say that all parties must make the same contribution to its management. According to Sands, the principle of common but differentiated responsibilities of states entails two elements: 'the common responsibility of all States for certain international issues, and differences in the extent of their international obligations to respond to those issues.'[19] Application of the principle of common but differentiated responsibilities could facilitate the international support, especially for developing countries, that is needed for the immense task of implementing the right of access to clean water and adequate sanitation for all people.[20] According to the Rapporteur of the ILA Committee on Legal Aspects of Sustainable Development, the principle of common but differentiated responsibilities 'has a firm status in various fields of international law, including human rights law, international trade law and international environmental law.'[21]

The granting of differential and more favourable treatment of developing countries, for example, is allowed under the GATT by means of the principle of non-reciprocal treatment ("the Enabling Clause") and a num-

---

[17] UNGA Resolution 1803 (XVII), 14 December 1962, on Permanent Sovereignty over Natural Resources. Charter of Economic Rights and Duties of States, adopted by the UNGA on 12 December 1974. See also Schrijver (1997), pp. 308-311.

[18] ICJ, Judgement in the Case Concerning East Timor, Portugal v. Australia, *ICJ Rep.*, p. 102, para. 29. ICJ Advisory Opinion on *Legal Consequences of the Construction of a Wall in the Occupied Palestinian Territory*, para. 155.

[19] Sands (1995), p. 344.

[20] See on the principle of common but differentiated responsibilities in a development context, *e.g.*, Matsui (2004) and (2002).

[21] ILA Committee on Legal Aspects of Sustainable Development (2002), p. 9.

ber of other provisions, in particular Article XVIII and Part IV.[22] In the 1992 Climate Change Convention, common but differentiated responsibilities forms an important principle. Article 3(1) of the UNFCCC states:

> The Parties should protect the climate system for the benefit of present and future generations of humankind, on the basis of equity and in accordance with their common but differentiated responsibilities and respective capabilities. Accordingly, the developed country Parties should take the lead in combating climate change and the adverse effects thereof.

The principle of common but differentiated responsibilities also has a prominent position in other documents that resulted from UNCED. Although the Watercourses Convention does not include a provision on common but differentiated responsibilities, in its preamble the situation of developing countries is emphasised, recalling the principles and recommendations adopted by UNCED in the Rio Declaration and Agenda 21. Principle 7 of the Rio Declaration formulates the principle as follows.

> States shall cooperate in a spirit of global partnership to conserve, protect and restore the health and integrity of the Earth's ecosystem. In view of the different contributions to global environmental degradation, States have common but differentiated responsibilities. The developed countries acknowledge the responsibility that they bear in the international pursuit of sustainable development in view of the pressures their societies place on the global environment and of the technologies and financial resources they command.

The WSSD Plan of Implementation refers more than once to the principle of common but differentiated responsibilities as set out in principle 7 of the Rio Declaration, for example, within the context of changing unsustainable patterns of consumption and production in which developed countries are to take the lead.[23] In this Plan, it is emphasised that, although each country has the primary responsibility for its own sustainable development and poverty eradication, concerted and concrete measures are required at all levels to enable developing countries to achieve their sustainable development goals.[24] These measures therefore include those needed to decrease the proportion of people without access to safe drinking water and basic sanitation.

---

[22] The Enabling Clause is officially named Differential and More Favourable Treatment, Reciprocity and Fuller Participation of Developing Countries, adopted at Tokyo (BISD 26S/203), Decision of 28 November 1979 (L/4903). See Section 5.4.2 of this book on the WTO/GATT.

[23] World Summit on Sustainable Development, Plan of Implementation, para. 13.

[24] World Summit on Sustainable Development, Plan of Implementation, para. 6.

Principle 3 of the ILA New Delhi Declaration is arguably the most detailed elaboration of the principle of common but differentiated responsibilities and its implications.[25] It points out not only the common but differentiated responsibilities of states, but also of other relevant actors. States, intergovernmental organizations, NGOs, corporations and civil society all need to cooperate in the achievement of global sustainable development. The basis for differentiation of responsibilities is twofold: the contribution of a state to the emergence of environmental problems and the economic and developmental situation of a state. Special regard is to be given to the needs and interests of least developed countries, while developed countries should take on primary responsibility in issues relating to sustainable development.

7.3   Social-ecological: life support by water

Life support by water is identified as the over-arching concept capable of responding to the conflict of interests between the social and the ecological pillars of sustainable development: *utilisation vs. conservation.*

At the community level, this conflict can be expressed as demand *vs.* availability. A large part of the population uses water wastefully while the other part is in real need of more water. Available water resources are declining, not in amount as such but in terms of resources that are readily and sufficiently accessible. According to the WWDR, it is expected that available drinking water will on average decline by about 30 per cent per capita due to a decline in availability coupled with an almost uncontrolled rise in demand. If water resources are to be sustainable, thinking must be oriented toward availability and essential needs rather than wants. The desires of individuals for water do not translate into a collective vision on the needs of present and future generations.[26] Increased cooperation and sharing of the burdens and benefits of water may encourage consensus on common interests.

At the national level, the conflict of interests can be expressed by sovereignty *vs.* integrity. Sovereignty of states in principle allows states to decide their own developmental and environmental use of freshwater resources in their territory. Territorial integrity of states can, for example, imply that the flow of a river may not be disturbed by an upstream state. In Chapter 3 it was concluded that absolute territorial sovereignty and absolute territorial integrity do not provide realistic descriptions of the actual state of affairs in the allocation of water, and that the doctrine of limited

---

[25] See www.un.org/ga/57/document.htm for the ILA New Delhi Declaration.

[26] Rousseau argues that law is to be guided by the common will as objectively and reasonably desirable for the community as a whole, not by the actual will or desire of the individuals. In order to respect the human dignity, however, this should not result in a disregard of the individual will.

territorial sovereignty and integrity comes closest to state practice. Limited sovereignty does combine the principles of sovereignty and integrity, but does not necessarily entirely overcome the conflicts of interests. For example, it may not include the interests of non-riparians in the sustainable development of an international watercourse within the context of the global water systems. Community of interests, on the other hand, does seem to provide a concept capable of overcoming the contradiction. As stated by McCaffrey:[27]

> The constant movement of the Earth's water through the hydrological cycle means that it would be futile for any one state to attempt to subject freshwater within its borders to absolute control. It also means, however, that the international community has a strong interest in this resource, including its protection and equitable apportionment. It would be going too far in the current state of international law to suggest that all freshwater is *res communis*. But it is critical that states begin to conceive of the *hydrologic cycle* in this way.

At the international level, the dilemma can be formulated as population *vs.* ecosystems. By its very existence, humankind is bound to alter the environment. The basic needs of a population already require the utilisation of natural resources such as water. In many of its aspects, however, this current utilisation shows enormous waste and a lack of awareness at best of the effects on both environment and people.[28] Sustainable development of the earth demands conservation of its waters, at least to a certain extent. The possibilities for such conservation would seem to lie in a combination of controlling population growth and the demands of that population, together with technical advances as well as more considerate handling of freshwater resources.

Life support by water can be translated into the following international law principles as a response to the dilemmas faced at the community, national and international levels: the right to a healthy environment; the precautionary principle; and eco-justice.

7.3.1   The right to a healthy environment

It has become widely acknowledged that a healthy environment is required for human health and development.[29] It may be clear from previous chap-

---

[27] McCaffrey (2001), p. 53.

[28] Freestone (1999), p. 364, having referred to effects of deforestation and overfishing: 'Governments are only beginning to appreciate the inherent value of their natural resources. The challenge of sustainable development is the internalization of these values into national resource assessment and decision making.'

[29] That human well-being and environment can progress beyond conflicting interests, and even enter into a relation of mutual assistance, may be illustrated by Norwegian programmes which encouraged the consumption of grains and vegetables rather than

ters that a healthy environment is at least a necessary condition for guaranteeing human rights such as the rights to life, health, and an adequate standard of living. The interrelationship between human rights and the environment is also increasingly recognised, for example, by the 1994 Draft UN Principles on Human Rights and the Environment of the UN Sub-Commission on Minorities.[30] The UN Commission on Human Rights states: 'the promotion of an environmentally healthy world contributes to the protection of human rights'.[31] Principle 1 of the Stockholm Declaration formulates the right to a healthy environment as follows:[32]

> Man has the fundamental right to freedom, equality and adequate conditions of life, in an environment of quality that permits a life of dignity and well-being, and he bears a solemn responsibility to protect and improve the environment for present and future generations.

The question remains, however, whether a distinctive right to a healthy environment is embedded within international law and if such a right could be classified as a human right. On this question, there is a debate currently taking place among various authors.[33]

The Universal Declaration and the 1966 human rights conventions do not include a human right to a healthy environment. The Human Rights Committee has dealt with environmental considerations but, even though it stressed the link between human rights and the environment, it did not qualify the right to a healthy environment as a human right either.[34] Neither does, for example, the Convention on the rights of the child, although

---

fattier grain-fed meats, both promoting the health of the population and reducing energy consumption, as cited by Schachter (1977), p. 141.

[30] UN Commission on Human Rights, Sub-Commission on Prevention of Discrimination and Protection of Minorities, Human Rights and the Environment, Final Report of the Special Rapporteur, *UN Doc*. E/CN.4/Sub.2/1994/9 (6 July 1994), 74.

[31] Resolution 1995/14 of the UN Commission on Human Rights, *UN Doc*. E/CN.4/RES/1995/14.

[32] 1972 Declaration of the UN Conference on the Human Environment (Stockholm Declaration).

[33] On a right to a healthy environment, see Giorgetta (2002) who argues in favour of such a right and, at p. 181, states that: 'the right to a healthy environment lends itself to immediate implementation by various bodies, under existing mechanisms for enforcing and/or interpreting regional and international human rights instruments and national constitutional provisions.' See also Sachs (1995), referring at p. 47, to moral consensus upon a right to freedom from environmental degradation, and, at p. 54, to a universal right to a healthy environment. Boyle (1996), pp. 48-65, questions the need for a human right to the environment in international law.

[34] See Giorgetta (2002), p. 177, on cases of the Committee dealing with environmental considerations.

reference is made to the dangers and risks of environmental pollution.[35] The African Charter on Human and Peoples' Rights does, however, include a right of peoples to a healthy environment: 'All peoples shall have the right to a general satisfactory environment favorable to their development.'[36] A remarkable but also quite unique reflection of a human right to a healthy environment, integrated with the right to life, can be found in Article 11 of the 1988 Protocol of San Salvador: '1. Everyone shall have the right to live in a healthy environment and to have access to basic public services. 2. The States Parties shall promote the protection, preservation and improvement of the environment.'[37] The ECE Aarhus Convention strongly supports a right to a healthy environment, stating in Article 1:

In order to contribute to the protection of the right of every person of present and future generations to live in an environment adequate to his or her health and well-being, each Party shall guarantee the rights of access to information, public participation in decision-making, and access to justice in environmental matters in accordance with the provisions of this Convention.

On the one hand, there are strong indications to support a right to a healthy environment in addition to the regional treaties. According to Judge Weeramantry, in his Separate Opinion to the *Gabcíkovo-Nagymaros* case, environmental rights are human rights.[38] In the *Minors Oposa* case, the Philippine Supreme Court stated that the 'right to a sound environment' of the minors also poses a duty to guarantee that right in their turn for future generations.[39] According to the Legal Expert Group of the Brundtland Commission: 'All human beings have the fundamental right to an environment adequate for their health and well-being.'[40] Principle 1 of

---

[35] CRC Article 24 (2c) on measures to combat disease and malnutrition in order to implement the right of the child to the enjoyment of the highest attainable standard of health.

[36] Article 24 of the African Charter of Human and Peoples' Rights.

[37] Article 11 of the 1988 Additional Protocol to the American Convention on Human Rights in the Area of Economic, Social, and Cultural Rights, San Salvador, entry into force: November 1999, 28 *ILM* (1989) 698.

[38] Separate Opinion to the ICJ Judgment in the *Gabcíkovo-Nagymaros* case, under B (b).

[39] Supreme Court of the Philippines, *Minors Oposa v. Secretary of the Department of Environment and Natural Resources* (Juan Antonio Oposa and others v. The Honourable Fulgencio S. Factoran and others), case of 30 July 1993, 33 *ILM* (1994), 173. See on this case also Section 4.4 of this study.

[40] Expert Group on Environmental Law of the World Commission on Environment and Development (1987), Annexe 1, Summary of Proposed Legal Principles for Environmental Protection and Sustainable Development Adopted by the WCED Experts Group on Environmental Law, I. General Principles, Rights, and Responsibilities, para. 1, titled Fundamental Human Right, p. 347.

the Rio Declaration states: '[human beings] are entitled to a healthy and productive life in harmony with nature.'

On the other hand, court cases recognising the right to a healthy environment are not overwhelming. Moreover, the international instruments including a right to a healthy environment, although authoritative, are either not legally binding or of only regional application, and mostly do not explicitly refer to a distinctive human right. The existence of a human right to a healthy environment remains controversial and is not firmly embedded in customary international law.

However, the right to a healthy environment is needed for the implementation of many other rights, including human rights; it is increasingly acknowledged as a right of people; it does appear to constitute a regional human right; and the contours can be identified of an emerging separate international human right to a healthy environment. If such a right is established, it would provide an additional way to protect both ecological and social interests at the community level.

7.3.2   The precautionary principle

In the application of the precautionary principle, a certain degree of scientific uncertainty is no argument against acting on an obligation to take measures when there is a risk of serious or irreversible damage to the environment.[41] The principle needs to be applied to prevent possible water-related detrimental effects on public health and environment, reflecting the bio-, geo- and chemical complexity of water, over which there may exist scientific uncertainty. The precautionary principle is stated in Article 15 of the Rio Declaration as follows:[42]

> In order to protect the environment, the precautionary approach shall be widely applied by States according to their capabilities. Where there are threats of serious or irreversible damage, lack of full scientific certainty shall not be used as a reason for postponing cost-effective measures to prevent environmental degradation.

Principle 4 of the ILA New Delhi Declaration relates the precautionary approach to human health, natural resources and ecosystems.[43] In case of possible significant harm, the precautionary approach is stated to be applicable to states and non-state actors. Principle 4 moreover includes accountability for harm caused, consideration in an EIA of all alternatives to achieve an objective, and a shift in the burden of proof in the case of possi-

---

[41] See Hunter, Salzman and Zaelke (2002), pp. 405-411, and Birnie and Boyle (2002), pp. 115-121.
[42] Reference is made to the precautionary approach instead of principle because of US insistence, see Birnie and Boyle (2002), p. 116.
[43] See www.un.org/ga/57/document.htm.

ble serious long-term or irreversible harm.[44] The Principle also states that decision-making processes should always support a precautionary approach to risk management and that precautionary measures should be based on independent scientific judgment, be transparent and not result in economic protectionism. The classification "significant" of the harm corresponds with the classification used for the no-harm principle.[45]

Although still debated, it appears that the precautionary principle is becoming more and more established in international environmental law.[46] The logic behind this principle is that dealing with threats such as global climate change with clear scientific indications of serious and irreversible damage cannot wait for full scientific certainty. The Berlin Rules underline the importance of the precautionary principle for the protection of aquatic environments (Article 23) and groundwater (Article 38). Like climate change, the long-term effect of the pollution of groundwater may be uncertain, but if this occurs it can be both detrimental and irreversible. In anticipating such effects, the precautionary approach to water use is essential, not only to safeguard the environment but also human health and economy. Taking further into account the lack of information on complex interactions and groundwater resources, the precautionary approach would seem to be an appropriate principle to apply to water allocation in case significant harm may result.

7.3.3   Eco-justice

The combination of respect for both human beings – including their social and economic interests – and for the environment can be expressed through environmental justice, formulated by Sachs as eco-justice.[47] This is not an established principle of international law but its ingredients are part of international law.

The need for environmental and human rights activists to combine forces can be illustrated by the struggle against Amazone deforestation and for the rights of its inhabitants, by the 1984 Bhopal gas leak that caused the death of thousands of people, and by the execution in Nigeria of Ken Saro Wiwa in 1995.[48] The interdependence between human rights and a healthy

---

[44] See Section 6.2.3 on environmental impact assessment in general.

[45] The no-harm principle is categorised as the principle bridging the economic and ecological pillars at the national level and therefore elaborated upon in Section 7.4.2.

[46] The ILA Committee on Water Resources Law (2004), at p. 28, states that the precautionary principle is included in almost all international environmental instruments adopted since 1990 and in the ILA New Delhi Declaration, but that it is not mentioned by the ICJ in the *Gabcíkovo-Nagymaros* case and is explicitly embraced as a legal obligation only by the Indian and Pakistan Supreme Courts.

[47] On eco-justice, see Sachs (1995). See also Picolotti and Taillant (2003) and Boyle and Anderson (1996).

[48] On the Amazone, see Sachs (1995), pp. 5-6, who also cites Amnesty International on the fact that in the 1980s over a thousand land-related murders were committed in

environment can moreover be illustrated by the discussion of the right to water and the right to a healthy environment earlier in this book. According to Sachs:[49]

> In the end, environmental justice is such a powerful concept because it brings everyone to the same level – that of shared dependence on an intact, healthy environment. The potential coalition surrounding environmental justice issues, in other words, is immense: everyone is willing to fight for something like clean water.

For the protection of both human rights and a healthy ecology, permanent monitoring of circumstances and applicable norms and standards is required. As stated by the ICJ in the *Gabcíkovo-Nagymaros* case: 'Such new norms have to be taken into consideration, and such new standards given proper weight, not only when States contemplate new activities but also when continuing with activities begun in the past.'[50] This stimulates eco-justice to be part of existing and new state activities, for example, by offering a way for modern environmental concerns to be integrated in the application of principles of justice. Products such as biological fair trade coffee provide another example of combining social and ecological interests.[51]

7.4   Economic-ecological: sustainable use of water

Sustainable use of water is identified as the over-arching concept placed in the foreground in response to the conflicting needs of the economic and the ecological pillar of sustainable development: *freedom vs. responsibility.*[52]

At the community level, this conflict can be expressed as rights *vs.* duties. Although these two elements may appear to be in conflict, the one cannot actually be defined without the other. That duties are the counter-

---

rural Brazil, resulting in less than ten convictions. On the Bhopal accident, see www.bhopal.com/review.htm and www.bhopal.net. See on the Ogoni case and death of Ken Saro Wiwa, *e.g.*, archive.greenpeace.org/comms/ken/murder.html.

[49] Sachs (1995), pp. 53-54.

[50] *Gabcíkovo-Nagymaros* case, ICJ Reports 1997, para. 140, p. 67.

[51] Fair Trade Federation (FTF) criteria include paying a fair wage in the local context and engagement in environmentally sustainable practices, see www.fairtradefederation.com.

[52] On sustainable use of fresh water, see Holland, Blood and Shaffer (2003), Hey (1995) and IUCN, UNEP and WWF (1991), Chapter 15. As stated earlier in Sections 1.3 and 5.2.3 of this study, sustainability refers to notions such as long-term considerations and common responsibility and is an important part of but not the same as sustainable development.

part of rights seems acknowledged, for example, in Article 1 of the UNGA Declaration on Social Progress and Development of 11 December 1969:

> All peoples and all human beings without distinction as to race, colour, sex, language, religion, nationality, ethnic origin, family or social status, or political or other conviction shall have the right to live in dignity and freedom and to enjoy the fruits of social progress and should, on their part, contribute to it.

Responsibility as the counterpart of freedom may be easily disregarded but may also be a core element of the path to sustainable development. Empowerment of people could provide a means by which to emphasise their part in the structure known as the international community.

At the national level, the problem can be expressed as independence *vs.* interrelationship, focussed here on the independence of and interrelationship between states. As with individual autonomy, the nation-state must be acknowledged, both in its identity and its freedom to decide on the basis of its own beliefs. According to French:[53]

> In the attempts to codify and progressively develop the present state and future direction of international law in the field of sustainable development, it is imperative that the international community continues to uphold public governance as being of pivotal importance in the attainment of balanced and global sustainable development. Without this, the globalizing society faces a much more uncertain, insecure and inequitable future.

The growing interdependence with other states and the world as a whole, however, marks the limits to this national freedom. Interrelationship is the framework within which states will have to preserve their identity and relative independence.

At the international level, this conflict of interests can be expressed, by analogy to the law of the sea, by *aquae liberum vs. aquae clausum*. On the one hand, freedom of water appears an attractive concept in acknowledging the dynamics and other natural characteristics of water. According to Grotius:[54]

> Thus a river, viewed as a stream, is the property of the people through whose territory it flows, or of the ruler under whose sway that people is... [T]he same river, viewed as running water, has remained common property, so that any one may drink or draw water from it.

---

[53] French (2002), p. 146.
[54] 'Of Things which belong to Men in Common' in *De juri belli ac pacis* (1625), Grotius, Lib. II, Cap II, XII, p. 196, as quoted in McCaffrey (2001), p. 150.

On the other hand, a certain control over water is needed to protect the common interests involved in water, considering that freedom often was – and in many cases may still be – a right of the most powerful, *cf.* the interests of colonial powers when advocating the freedom of the high seas. Moreover, regulation is needed to address global problems such as acid rain, climate change and overfishing, limiting freedom that can result in a tragedy of the commons. Under proper conditions, regarding the world's water as a global common good can serve the common interests. For example, freedom of navigation was used by the colonial powers to strengthen their positions but can also provide landlocked states with access to the sea, as in the case of SADC.[55] In search for a regime that balances freedom with control for fresh water, parallels could be drawn with the law of the sea.

Sustainable use of water can be translated into the following international law principles as a response to the dilemmas faced at the community, national and international level respectively: the polluter and user pays principle; the no-harm principle; and the common heritage or concern of humankind.

7.4.1   The polluter and user pays principle

The polluter and user pays principle provides an incentive for polluters and users to reduce, respectively, their pollution and consumption.[56] The status of the polluter and user pays principle is controversial and cannot be said to be an established principle of international law. It does, however, at least provide a guiding instrument for effective policy, also partly enabling the wider implementation of the no-harm principle, since pollution is frequently caused by private parties. Principle 16 of the Rio Declaration states:

> National authorities should endeavour to promote the internalization of environmental costs and the use of economic instruments, taking into account the approach that the polluter should, in principle, bear the cost of pollution, with due regard to the public interest and without distorting international trade and investment.

The internalisation of social and environmental externalities in prices of products and uses improves the functioning of the market mechanisms and could result in reallocation of, for example, water, toward more sustainable uses. State intervention is required to implement the principle at the com-

---

[55] See on the SADC, *e.g.*, Salman (2001a).
[56] The polluter pays principle was first found in the Recommendation of the OECD Council on Guiding Principles Concerning International Economic Aspects of Environmental Policies, Annex I, adopted at its 239th meeting, May 26, 1972. See Hunter *et al.* (2002), pp. 412-414.

munity and national level. Most countries do not apply the polluter and user pays principle, including the industrialised countries that would be expected to have the means to implement it. For developing countries, the efforts needed for such internalisation provide a barrier, plus the fact that standardisation of environmental criteria could damage their competitiveness.

Making consumers pay a price that reflects both the costs of pollution and factors such as the need for a fair price for farmers in the case of agricultural products could encourage a change in the use of water. Sustainable production of food could be further stimulated by reallocating subsidies paid for unsustainable agricultural practices toward, for example, income guarantees and sustainable products. Water use, as well as pollution, can moreover be reduced by changing to lower meat consumption.[57] There might be cases in which the polluter and user pays principle is not an appropriate instrument. When serious environmental damage occurs, for example, penalties under criminal law could be found to be more appropriate. Moreover, pollution is not the only harmful activity of humans.

7.4.2   The no-harm principle

The no-harm principle is firmly embedded within international law and can be regarded as an expression of principles of good neighbourliness or *sic utere tuo, ut alienum non laedes*.[58] In the *Chorzow Factory* case, the PCIJ stated that: 'It is a principle of international law, and even a general conception of law, that any breach of an engagement involves an obligation to make reparation.'[59] In the *Trail Smelter* Arbitration, Canada was found to be responsible for damage on USA territory, providing for a penalty in case of continuance.[60] The ICJ reiterated state responsibility for a breach of an international obligation, leading to a duty to pay compensation, in the *Corfu Channel* case.[61] According to the ICJ, following the statement quoted earlier on the environment not being an abstraction, in the *Legality of the Threat of Use of Nuclear Weapons* case: 'The existence of the general obligation of States to ensure that activities within their jurisdiction and control respect the environment of other States or of areas beyond national control is part of the corpus of customary international law relating to the

---

[57] See www.profetas.nl on research of ways to partially replace animal proteins with plant proteins.

[58] See Nelissen (2002), Birnie and Boyle (2002), p. 104, and McCaffrey (2001), pp. 349-353. On environmental harm, see Birnie and Boyle (2002), pp. 104-137.

[59] *Case Concerning The Factory at Chorzow* (Claim for Indemnity) (The Merits), PCIJ Ser. A, No. 17 (1928).

[60] *Trail Smelter Arbitration* (USA *v*. Canada), first decision in 1938, *AJIL* 33 (1939) 182, and second decision in 1941, *AJIL* 35(1941), 684.

[61] *Corfu Channel* case (UK *vs*. Albania), Judgment of 9 April 1949, ICJ Rep. (1949).

environment'.[62] The field of application of the no-harm principle thus includes the territory of other states and certain areas of an international status such as Antarctica, the high seas and outer space.[63]

Within international water law, the no-harm principle can be found in, for example, the ECE Convention Article 2, which obliges parties to 'take all appropriate measures to prevent, control and reduce any transboundary impact'. The definition of transboundary impact in the ECE Convention implies 'any significant adverse effect on the environment resulting from a change in the conditions of transboundary waters caused by a human activity', in which "effect on the environment" is broadly defined.

Under the Watercourses Convention, parties are obliged to prevent the causing of significant harm and to take actions whenever such harm is nonetheless caused. Article 7 of the Watercourses Convention, on the obligation not to cause significant harm, states:

> 1. Watercourse States shall, in utilizing an international watercourse in their territories, take all appropriate measures to prevent the causing of significant harm to other watercourse States.
> 2. Where significant harm nevertheless is caused to another watercourse State, the States whose use causes such harm shall, in the absence of agreement to such use, take all appropriate measures, having due regard for the provisions of Articles 5 and 6, in consultation with the affected State, to eliminate or mitigate such harm and, where appropriate, to discuss the question of compensation.

The no-harm principle is further emphasised by Principle 21 of the Stockholm Declaration and Principle 2 of the Rio Declaration, limiting the sovereign right of a state to exploit their own resources, pursuant to their own environmental and developmental policies, by the condition that no harm must be caused beyond their jurisdiction.

The absolute form of the no-harm principle would be similar to absolute territorial integrity, resulting in a regulation as inequitable as absolute territorial sovereignty. Allowing states to cause serious or irreversible harm beyond their territory would be an application of absolute territorial sovereignty. For example, Ethiopia could develop and act in any way it wishes

---

[62] ICJ Advisory Opinion on the *Legality of the Threat of Use of Nuclear Weapons in Armed Conflicts*, para. 29. See also the Trail Smelter Arbitration between the USA and Canada: 'under the principle of international law as well as the law of the United States, no state has the right to use or permit to use its territory when the case is of serious consequences and the injury is established by clear and convincing evidence.'; The Corfu Channel case: 'it is every state's obligation not to allow knowingly its territory to be used for acts contrary to the rights of other states.' the Lake Lanoux Arbitration: 'interdiction prohibiting a state upstream to alter the water of a river in such condition as to cause substantial damage to the downstream states.'

[63] The duty of states to minimise environmental harm as formulated in Article 8 of the Berlin Rules does not require a transboundary setting.

even though that would lead to a use of the Nile causing the destruction of the Nile-based economy of Egypt.[64] In these days of over-exploitation, an obligation of a state to cause absolutely no harm beyond its territory would be an unworkable application of absolute territorial territory. For example, Ethiopia would then not be allowed to develop, since every action would cause some change in the flow of the Nile entering Egypt. The necessary balance is reflected in the qualification "significant" harm, which is discussed next.

The reasonable use of property and territory can under some circumstances cause factual harm, but that does not in itself lead to legal harm. According to McCaffrey: 'To be sure, the causing of some forms of harm may be considered per se unreasonable, as, for example, where the harm endangers human health or is of an irreparable or long-lasting nature.'[65] The final text of the Watercourses Convention mentions "significant" harm, while the Draft Articles referred to "appreciable" harm.[66] The Helsinki Rules apply to "substantial" pollution.[67] The dominant term nowadays, used both by the ECE and Watercourses conventions, is "significant", which is considered to be more than "trivial" and less than "substantial" or "serious" harm.[68] Its actual content is to be determined case by case, depending on the circumstances.

Harm caused in relation to freshwater resources may concern either water quantity and/or quality, such as water flow and pollution, which are normally interconnected. In view of such aspects as the unknown flow of many underground aquifers and the likely delayed disclosure of such harm, the establishment of harm, let alone its degree, in the case of groundwater is likely to be a difficult task, underlining the importance of applying the precautionary approach and a thorough impact assessment.

The obligation not to cause significant harm, as formulated in the ECE and Watercourses conventions, is one of due diligence. According to McCaffrey: 'Procedurally, if a state makes a prima facie showing that it has been significantly harmed as a result of another state's conduct in relation to an international watercourse, the burden in effect shifts to the alleged source state to show that it has fulfilled its obligation of due diligence to prevent the harm.'[69] McCaffrey continues by observing that state responsibility results when the harming state is incapable of showing due diligence, and that even if the duty of due diligence is met, the conduct must also be

---

[64] See Wiebe (2001) for a discussion on the Nile, including the Nile Basin Initiative. On the Nile, Niger and Senegal river systems, see Godana (1985).
[65] McCaffrey (2001), p. 365.
[66] Report of the International Law Commission on the Work of its Forty-Sixth Session, UN Doc. A/49/10 (1994), p. 236.
[67] Article X of the 1966 ILA Helsinki Rules.
[68] McCaffrey (2001), pp. 369-370.
[69] McCaffrey (2001), p. 380.

in line with equitable and reasonable utilization.[70] Such harm caused by a state would indicate that the balance of interests has tipped too far. As formulated by Tanzi and Arcari, discussing the Watercourses Convention: 'Setting out the no harm rule in a separate article gives it priority over the other individual factors. In functional terms, this priority implies the presumption of the inequitable character of a use that causes significant harm.'[71] They also conclude that in such cases the burden of proof shifts to the state that is causing significant harm.[72]

7.4.3   Common heritage or concern of humankind

Recognising global freshwater resources as a common heritage of humankind, or at least as a common concern of humankind, would further underline the interrelationship between water resources and the need for their use to be sustainable, possibly reflected, for example, by eco-labelling.[73]

The principle of the common heritage of humankind reflects the special position of global commons and requires management to take the global interests involved into account, through such concepts as non-appropriation, international management, shared benefits and use for peaceful purposes.[74] According to Brown Weiss, the doctrine of common heritage anticipates the need for planetary obligations.[75] Referring to Ambassador Pardo of Malta, she mentions non-ownership of the heritage, shared management, shared benefits, usage exclusively for peaceful purposes, and conservation for mankind as the five principal elements of common heritage.[76] The principle of the common heritage of humankind could promote the involvement of the international community in its efforts to eradicate poverty and to protect the environment. Apart from exceptional cases regulated in treaties with respect to areas beyond national jurisdiction such as the deep seabed and the moon, the common heritage of humankind can at most be said to be an emerging principle of international law.

The principle of the common heritage of humankind can be found in the 1982 UN Convention on the Law of the Sea (UNCLOS), applied there to the deep seabed, as well as in the 1979 Agreement Governing the Activi-

---

[70] *Ibid.* On the relation between the principle of equitable and reasonable utilization and the no-harm principle, see Section 3.3.2 of this study.

[71] Tanzi and Arcari (2001), p. 179; they continue to state that 'This presumption could still be challenged in relation to other factors, "with special regard being given to the requirements of vital human needs", under Article 10(2).'

[72] *Ibid.* They add that the ways and means offered by the harming state are to be considered in good faith by the victim state.

[73] On sustainability labelling, see Campins-Eritja and Gupta (2002).

[74] See Hunter *et al.* (2002), pp. 392-393.

[75] Brown Weiss (1989), p. 48.

[76] Brown Weiss (1989), pp. 48-49. For further elaboration of the intergenerational perspective of global commons, see Brown Weiss (1989), pp. 289-291.

ties of States on the Moon and Other Celestial Bodies, concerning for example outer space.[77] At the regional level, the common heritage of mankind is referred to in Article 22 on the right to development of the African Charter on Human and Peoples' Rights. The principle of the common heritage of humankind is not incorporated in more recent treaties. Instead, in their preambles, the treaties on biodiversity and climate change refer to the common concern of humankind.

A shift from common heritage toward the principle of common concern of humankind can also be discerned in other documents. Principle 1.3 of the ILA New Delhi Declaration refers to the protection, preservation and enhancement of the natural environment as a common concern of humankind, while stating that the 'resources of outer space and celestial bodies and of the sea-bed, ocean floor and subsoil thereof beyond the limits of national jurisdiction are the common heritage of humankind.' Obviously, this is an effort to record the law as it currently stands.

According to the Rapporteur of the ILA Committee on Legal Aspects of Sustainable Development, the common concern of humankind 'is a somewhat vaguer notion than the common heritage principle, obviously not implying non-appropriation and an international regime, but it still carries the connotations of global interest in preserving the environment and needs of future generations.'[78] The principle of common concern gives the international community less legal grounds for intervening in the domestic affairs of countries, such as their implementation of access to water for their people. The status of the principle of common concern of humankind within international law appears stronger than that of the common heritage of humankind.

On the one hand, the global hydrological system is a suitable candidate to become a common heritage of humankind.[79] Besides international rivers that pass through different countries, water also travels across borders via the hydrological system, which includes lakes and groundwater. The hydrological system is a dynamic cyclical process, involving both seawater and fresh water.[80] Increased awareness of the implications of the hydrological cycle has resulted in careful steps toward its reflection in international law. For example, reference to the hydrologic system is made in

---

[77] Agreement Governing the Activities of States on the Moon and Other Celestial Bodies, adopted by the UN General Assembly in Resolution 34/68, New York, 5 December 1979, entry into force: 11 July 1984, 1363 *UNTS*, 3, and 18 *ILM* (1979), 1434. Status as of 4 October 2004: 11 parties and 11 signatories.

[78] ILA Committee on Legal Aspects of Sustainable Development (2002), p. 10.

[79] See Caflish (1992), p. 59, arguing in favour of internationalisation of shared natural resources situated beyond national jurisdiction, and Brown Weiss (1989), pp. 246-247, according to whom 'reasonable access and use of the natural resources of our common patrimony' constitutes planetary rights of communities.

[80] See Section 2.4.2 of this study on the hydrological cycle. See for an extensive elaboration on the hydrologic cycle and its implications for international law, McCaffrey (2001), Chapter 2.

Recommendation 51 of the UN Conference on the Human Environment, which emphasises that 'the net benefits of hydrologic regions common to more than one national jurisdiction are to be shared equitably by the nations affected'. For international law to function as a stimulus to sustainable development, it must increasingly take the hydrological system into account. Moreover, bringing the allocation of freshwater resources into line with sustainable development poses a complex task.[81] This means that the sustainable use of water is a global issue, which in many of its aspects cannot be managed by single states.[82] The importance to development of such an approach could lie in the sharing of the burdens and the benefits, more likely resulting in an equitable use of water than the allocation of water between countries with huge differences in means to exploit and manage them.[83] According to McCaffrey, the international community should considerably expand efforts to alleviate water shortages, referring to its stewardship of the hydrologic cycle.[84] He goes on to state:[85]

> It is not difficult to find a factual predicate for such action. After all, most of the water that precipitates onto land evaporates from the ocean, and some two-thirds of the ocean lies beyond the limits of national jurisdiction. This portion of the sea, at least, is a 'commons'.

On the other hand, regarding freshwater resources as a common heritage within international law could be blocked by the notions of sovereignty and territoriality. It might furthermore lead to inequality of access to natural resources. It could create an obligation for developing countries that possess enough water to share those resources, while they have no access to other natural resources abundantly present in other countries. In view of the uncertainties inherent in the concept of a common heritage as applied to water allocation, combined with the modern tendency away from that concept, it seems not yet feasible for the international community to join forces in a way capable of addressing the hydrological cycle as a

---

[81] See, *e.g.*, Chapter 18 of Agenda 21. Sustainable use of fresh water implies, for example, that exploitation of water remains within recharge rates. Therefore, no mining of ground water aquifers should take place and their recharge areas are to be protected, see Brown Weiss (1989), p. 127. The importance of water resources for later generations is underlined by Brown Weiss (1989), pp. 232-247.

[82] Degradation of international waters is defined as one of the four critical threats to the global environment addressed by the Global Environment Facility (GEF) established in 1991, see www.gefweb.org.

[83] Gandhi (1992), p. 140, goes as far as to state: 'The exploitation of natural resources found in the common heritage of mankind is meant for the development of all countries in general.'

[84] Stewardship is also discussed by Barlow and Clarke (2002), pp. 211-213, and mentioned by Tarlock (1997), p. 16, who states: 'In short, sustainable development is enlightened, forward-looking resource stewardship.'

[85] McCaffrey (2001), p.53.

communality. At present, water as such is not often specifically classified as one of the issues that common heritage might extend to.[86]

7.5   Conclusions

In addition to earlier identified key concepts, this Chapter has identified the following key concepts that respectively bridge the social and economic, the social and ecological, and the economic and ecological pillars of sustainable development.

Development through water includes the right to development, the right of self-determination and the principle of common but differentiated responsibilities. The status of the right to development appears to be still controversial and apparently mired in an ideological discussion, while the right of peoples of self-determination can be argued to constitute a right *erga omnes*. The principle of common but differentiated responsibilities may require protection of the global environment by developed countries, while possibly allowing developing countries to focus on national concerns. On the one hand, it could therefore provide developing states with an instrument by which to prioritise the development of their own population, including their access to water when allocating water to different uses. On the other hand, the development of countries does not necessarily coincide with the right to development of peoples. A government could actually use the priority of the development of the state as an argument against the particular interest of a people. In the case of allocation of fresh water, however, such application of the principle of common but differentiated responsibilities could be a violation of international law, considering the requirement of governments to provide their people with an adequate standard of living and the right of self-determination of peoples.

Life support by water combines social and ecological interests through the right to a healthy environment, the precautionary principle and eco-justice. The right to a healthy environment, if genuinely emerging as a human right, has to be seen within the context of the other human rights and their indivisibility as underlined by the Vienna Declaration. The anthropo-centric orientation of such a right, and many other human rights for that matter, underlines the need to balance it with its counterpart, a responsibility to conserve the environment not only for others but also for the protec-

---

[86] The Rapporteur of the ILA Committee on Legal Aspects of Sustainable Development (2002), p. 10, mentions 'tropical rain forests, wetlands of international importance or the environment and what belongs to all of us, such as major ecological systems of our planet' as examples of such fields. Through wetlands and the last category, the hydro-logical cycle could be included. Water is mentioned within the context of common patrimony by Brown Weiss (1989), pp. 289-291. Hohmann (1992), p. 279, argues that common heritage to a certain degree also has to be respected with regard to wetlands of international importance (Ramsar Convention).

tion of the ecosystem itself. The right to a healthy environment could become part of international development law, international environmental law and human rights; in other words, it could be a typical principle of sustainable development. The precautionary principle and eco-justice further point out the sustenance of all life by water.

The polluter and user pays principle, the no-harm principle and the common heritage or concern of humankind can contribute to *sustainable use of water*. Environmental costs can be internalised by means of the polluter and user pays principle. In the case of serious or irreversible harm, the status of an affected state and the environment can be improved by a shift in the burden of proof. In due course, the increased urgency of the need to deal with water in a sustainable manner, together with such factors as the increased international cooperation resulting from globalisation, might induce the international community to regard the hydrological cycle as a common heritage of humankind. At present, the common concern of humankind is better established within international law and probably best serves the recognition of water as a global good. It emphasises the shared responsibility of the international community for the provision of access to water for all and the protection of ecosystems. Reinforcement of the common concern of humankind could balance economic rights with ecological duties. Following the lead of Grotius' *Mare Liberum* (freedom of the sea), the time may now have come for a conditional *Aquae Liberum* (freedom of water).[87] Freedom within this context would include a freedom of water itself, such as freedom from unsustainable use, and entails rights and duties of humankind. It would allow for human use of water, but also require the world's water to be protected in the common interest by humankind in its role of guardian.[88] Balancing the central role of human interests with their corresponding responsibilities seems to be the key to sustainable development of water.

---

[87] 'So by the decree of divine justice it was brought about that one people should supply the needs of another, in order, as Pliny the Roman writer says, that in this way, whatever has been produced anywhere should seem to have been destined for all.', Hugo Grotius, Chapter I of *The Freedom of the Seas or The Right Which Belongs to the Dutch to take Part in the East Indian Trade*, translated by R. Van Deman Magoffin, Oxford University Press: New York, 1916, as reproduced in L.E. van Holk and C.G. Roelofsen (eds) (1983), *Grotius Reader: A reader for students of international law and legal history*, T.M.C. Asser Institute: The Hague, p.59.

[88] The term guardian is also employed by Brown Weiss (1989), who at p. 109, within the context of the enforcement of planetary rights, argues in favour of the state serving as a guardian *ad litem* for future generations and, at p. 123, furthermore argues in favour of standing in national courts and administrative bodies of 'a representative of future generations, who might function like a guardian ad litem.'

8.   Guardianship over water

8.1   Sustainable development of water

Throughout this book various existing and progressively developing principles of international law that can be instrumental to the achievement of sustainable development in water management have been identified and analysed. In Chapter 7 a first step was made to further review their relationship, resulting in the identification and analysis of principles that combine each set of pillars of sustainable development. In this concluding chapter, all three pillars of sustainable development are combined, resulting in an international law framework for sustainable development of water.[1] In order to stimulate the achievement of sustainable development in water management, the framework presented in this chapter aims to diminish the risk of trade-offs with a bias against any of the social, economic or ecological pillars.

In addition, the co-existence of abundance and wealth next to scarcity and poverty and the power play that results, can be an obstacle to cooperation in good faith at all policy-levels. For example, on the basis of past experiences, many developing countries fear that the environment might be protected at the cost of their development. Such developing countries are understandably cautious when it comes to sustainable development.[2] Nevertheless, some of the most progressive regional agreements are established in developing countries but their application carries a considerable financial burden as well as human and developmental costs. The developed countries, on the other hand, are less inclined toward such agreements but they do have the means to implement them. In order to balance the interests according to sustainable development, an enabling environment for cooperation in good faith is to be created.[3] It takes renewed commitment

---

[1] The sequence and categorisation of pillars, policy-levels, principles and concepts are reflected in the Content, in the methodology in Chapter 1, and the comprehensive framework in 8.3. To my knowledge, a comparable framework of international law does not yet exist.

[2] For example, Schwabach (1998), p. 279, concludes:
   'The trend in international watercourse law has turned away from absolute territorial sovereignty and toward protection of the rights of lower riparians. In concert with a general trend in international environmental law toward protecting the environment even at the expense of development, this trend has caused many developing countries to view international environmental law as an obstacle to development created by already-developed countries.'

[3] On North-South dilemmas in water management cooperation, see for example the EU-ACP Water Facility, europa.eu.int/comm/europeaid/projects/water/index_en.htm, in which case NGOs call for amendment of the current modalities to serve the interests of people in ACP countries confronted with water and sanitation problems, www.bothends.org/project/project_info.php?id=13&scr=tp.

to diminish the poverty gap within as well as between states and to, for example, reach the Millennium Development Goal to eradicate extreme poverty.[4]

Without denying the complexity of the issues to be overcome, it seems that the human mind tends to focus on differences rather than similarities. This functioning of the mind is very useful in categorising elements, and is indeed essential to our perception and comprehension of the world. Contradictions can also reveal, *e.g.*, arguments from various perspectives, variety in culture, and differences in definitions. Nevertheless, the focus on differences can conceal the fact that many apparent conflicts arise from different approaches to one and the same system. Analysing the contradictions can reveal a paradox, the study of which may lead to new insights that may actually combine the apparently incompatible interests.[5]

Considering that several elements are to be weighed in their comparative relation, the actual content of sustainable development of water will have to be found on a case by case basis. Barlow and Clarke in discussing the situation in the Kitlope valley, in northern British Columbia, provide an example of economically sustainable and ecologically sound management of a watershed – involving community, government and private actors – including the establishment of a training institute for people from the Aboriginal community to work in areas including guided tourism and ecosystem research.[6] These authors further state that recycling of industrial water has resulted in the stabilising of water use over two decades in western Germany.[7] Moreover, the eradication of poverty together with capacity-building can provide interesting new markets and partly remove the need for migration. To take another example, the provision of adequate drinking water will lead to fewer diseases and result in higher productivity, as will a healthy environment.

The key to surmounting apparent contradictions lies in acknowledging that delimitation and models are instruments for our comprehension of an interrelated world. Thus, development and environment are different aspects of the same coin of sustainable development. Moreover, both principles of justice and self-interest in the longer term provide reasons to foster sustainable development.[8] The common element surmounting the conflicting interests in the whole of sustainable development in water management

---

[4] See www.un.org/millenniumgoals for graphics on the implementation of the Millennium Declaration Goals and the 2003 Report of the Secretary-General on the Implementation of the United Nations Millennium Declaration, *UN Doc.* A/58/323.
[5] A paradox can be defined as a person or thing that combines contradictory features or qualities, Pearsall (1999), p. 1033. See also, *e.g.*, 'Paradoxen: Tegenstrijdigheden helpen bij helder denken', *Volkskrant* 28 februari 2004, 7W.
[6] See Barlow and Clarke (2002), p. 197.
[7] Barlow and Clarke (2002), p. 233.
[8] As expressed by Schachter (1977), p. 103-104: 'self-interest as a motivating force does not exclude a result which should be regarded as equitable because it meets pressing needs and legitimate expectation.'

can be expressed by the umbrella concept *guardianship over water*. Putting the model in perspective, it may be clear that the framework *Guardianship over Water* presents a hypothetical optimum. Elements of the framework can overlap and many of the principles are progressively developing but not yet firmly embedded within international law. In a way, the pillars are analogous to separate tributaries, flowing into one another without regard to human-made boundaries, either real or conceptual.

The identified principles of international law are now gradually further integrated into the framework of international law on guardianship over water. Thus, this chapter focuses on the process of systematically building a framework. For the textual elaboration of the principles and concepts, reference is made to earlier sections of this book. In constructing the framework, first, the combination of the principles within the pillars of sustainable development results in over-arching key concepts (8.2). Second, the combination gradually leads to the framework *Guardianship over Water* (8.3). Third, the operationalisation of the model is illustrated by a Draft Declaration on Guardianship over Water, a pricing mechanism and a preliminary assessment of the application of the framework to legal instruments (8.4).

8.2   Key concepts combining principles by pillars

The applicable principles of international law are now categorised for each pillar of sustainable development and subsequently integrated into a key 'concept'. In addition and considering the multidisciplinary elements of this study and related divergent views on the relevance of models and text among its target group, the principles are presented per policy-level as building-blocks of the framework in Annex I, including their underlying reason and a proposed example of a specifically appropriate tool for implementing the principle. The principles are linked to the policy-level to which they mainly apply, although it is acknowledged that to some extent all principles must be taken into account at all levels. The policy-making levels are classified as community, national and international levels.[9] At the community level, principles mainly relate to rights and duties of people. The principles categorised at the national level often include international aspects but mainly concern national policies of a state addressing its people or another state, while the principles at the international level address foremost the international community as a whole.[10]

---

[9] See for the categorisation in policy-levels Section 1.4.

[10] *Cf.* Birnie and Boyle (2002), p. 99:

'International lawyers have traditionally distinguished between legal obligations owed to another state, which can be enforced only by that state, and legal obligations owed to the whole international community of states, which can be enforced by or on behalf of that community.'

Social pillar: The social pillar at the community level entails a human right to water (4.2). At the national level, eradication of poverty (4.3) is the applicable social principle. Equity (4.4), intra- and intergenerational, is identified as part of the social pillar at the international level. Integrating the aforementioned social principles as they apply at all levels results in the key concept *access to water* (4.1).

Access to Water	Social pillar
Community level	human right to water
National level	eradication of poverty
International level	equity

Economic pillar: The economic pillar at the community level entails the right to use water (5.2); at the national level, water as an economic good (5) is the applicable economic principle; while a supportive and open international economic system (5.4) is identified as part of the economic pillar at the international level. Integration of the aforementioned economic principles as they apply at all levels results in the key concept *control over water* (5.1).

Control over water	Economic pillar
Community level	right to use water
National level	water as an economic good
International level	supportive and open international economic system

Ecological pillar: The ecological pillar at the community level entails the duty to protect water (6.2); at the national level, protection of the environment (6.3) is the applicable ecological principle; while ecological integrity (6.4) is identified as part of the ecological pillar at the international level. Integration of the aforementioned ecological principles at all levels results in the key concept *protection of water* (6.1).

Protection of Water	Ecological pillar
Community level	duty to protect water
National level	protection of the environment
International level	ecological integrity

Social and economic pillars: At the community level, the social and economic pillars entail the right to development (7.2.1); at the national level, the right of self-determination of peoples (7.2.2) is the applicable social and economic principle; while the principle of common but differentiated responsibilities (7.2.3) is identified as part of the social and economic pillars at the international level. Integration of the aforementioned social and economic principles as they apply at all levels results in the key concept *development through water* (7.2).

Development through Water	Social and economic pillars
Community level	right to development
National level	right of self-determination
International level	common but differentiated responsibilities

Social and ecological pillars: The social and ecological pillars at the community level entail the right to a healthy environment (7.3.1); at the national level, the precautionary principle (7.3.2) is the applicable social and ecological principle; while eco-justice (7.3.3) is identified as part of the social and ecological pillar at the international level. Integration of the aforementioned social and ecological principles at all levels gives us the key concept *life support by water* (7.3).

Life Support by Water	Social and ecological pillars
Community level	right to a healthy environment
National level	precautionary principle
International level	eco-justice

Economic and ecological pillars: The economic and ecological pillars at the community level entail the polluter and user pays principle (7.4.1); at the national level, the no-harm principle (7.4.2) is the applicable economic and ecological principle; while the common heritage or concern of humankind (7.4.3) is identified as part of the economic and ecological pillars at the international level. Integration of the aforementioned economic and ecological principles at all levels results in the key concept *sustainable use of water* (7.4).

Sustainable Use of Water	Economic and ecological pillars
Community level	polluter and user pays principle
National level	no-harm principle
International level	common heritage or concern of humankind

Social, economic and ecological pillars: The social, economic and ecological pillars at the community level entail the sum of human rights and duties (see also 3.2.1); at the national level, qualified sovereignty of states (see also 5.2.2) is the applicable social, economic and ecological principle; while the principle of equitable and reasonable utilization (see also 3.3.2) is identified as part of all pillars at the international level. The combination of the aforementioned social, economic as well as ecological principles at all levels results in the key concept *sustainable development of water* (8.1).

Sustainable Development of Water	Social, economic and ecological pillars
Community level	human rights and duties
National level	qualified sovereignty
International level	equitable and reasonable utilization

8.3   Framework on guardianship over water

The principles of international law can now be combined into a framework of international law on guardianship over water. The framework provides increased transparency to the interrelationship between the principles. They thus reveal the potential for balancing social, economic and ecological interests and provide guidance to decision-making in search for sustainable development in water management. The framework moreover aims at a more predictable outcome of the process of equitable and reasonable utilization aimed at sustainable development. First, a basic version of the framework on guardianship over water is presented, providing an overview of the key concepts and principles within the pillars of sustainable development. Second, a comprehensive version of the framework on guardianship over water is displayed.

The basic framework *Guardianship over Water*

Those concepts with relevance for one pillar are given first in row 1, followed by those in the fields of two pillars (in rows 2-4) and finally the one applying to all pillars in row 5.

Guardianship over Water		
Social Pillar	**Economic Pillar**	**Ecological Pillar**
Access to water • human right • eradication of poverty • equity	*Control over water* • right to use • economic good • open international economy	*Protection of water* • duty to protect • environmental protection • ecological integrity
Development through water • right to development • right of self-determination • common but differentiated responsibilities		
Life support... • right to a healthy environment • precautionary principle		*...by water* • ecological justice
	Sustainable use of water • polluter and user pays principle • no significant damage • common concern or heritage	
Sustainable development of water • human rights and duties • qualified sovereignty • adapted version of equitable and reasonable utilization		

The comprehensive framework *Guardianship over Water*

The comprehensive framework on guardianship over water provides an overview of the relationship between the combined pillars, policy-levels, key concepts, principles and their reasons and proposed examples of tools. Within each pillar and possible combination of pillars of sustainable development, key concepts integrate the identified principles of international law. The meaning of the key concepts at each policy-level is reflected by the principle mainly focussed on that level.

Each principle is furthermore placed in relation to its rationale – the reason that both underlies it and gives it direction. A "reason" provides the direction for regulation by a "principle", which in turn provides an instruction for implementation by means of a "tool". The reason and proposed example of a specifically appropriate tool for implementing the principle can also be found in Annex I. In addition to the specified tools, law can provide legal means at the various levels for the peaceful settlement of disputes which are basically identical for all principles: at the community level, recourse to administrative institutions or court; at the national level, implementation by legislation and through institutions; at the international level, maintaining of international relations through agreements such as treaties, guided by the rule of law and coordinated by joint bodies.[11]

---

[11] For an elaboration of international institutional law see, *e.g.*, Schermers and Blokker (2003) and Klabbers (2002).

Guardianship over Water		Social Pillar (Chapter 4)	Economic Pillar (Chapter 5)	Ecological Pillar (Chapter 6)
Key concept per pillar(s)		access to water (Section 4.1)	control over water (Section 5.1)	protection of water (Section 6.1)
Meaning of key concept per policy-level: o Principle of international law per focus area ▪ Reason per principle ▪ Tool per principle	o Community level ▪ Reason ▪ Tool	o human right to water ▪ basic human need ▪ reporting	o right to use water ▪ individual freedom ▪ community management	o duty to protect water ▪ collective responsibility ▪ community involvement
	o National level ▪ Reason ▪ Tool	o eradication of poverty ▪ solidarity ▪ capacity-building	o water as an economic good ▪ economic viability ▪ pricing	o protection of the environment ▪ degradation of nature ▪ environmental impact assessment
	o International level ▪ Reason ▪ Tool	o equity ▪ fairness ▪ access to information and to justice	o supportive and open international economic system ▪ long-term effectiveness ▪ common management	o ecological integrity ▪ interests in hydrological cycle ▪ catchment basin approach

Social and Economic Pillars (Chapter 7)	Social and Ecological Pillars (Chapter 7)	Economic and Ecological Pillars (Chapter 7)	Social, Economic and Ecological Pillars (Chapter 8)
development through water (Section 7.2)	life support by water (Section 7.3)	sustainable use of water (Section 7.4)	sustainable development of water (Section 8.1)
o right to development	o right to a healthy environment	o polluter and user pays principle	o human rights and duties
▪ quality of human life	▪ quality of living area	▪ individual responsibility	▪ human dignity
▪ education	▪ price differentiation	▪ internalisation of externalities	▪ subsidiarity
o right of self-determination	o precautionary principle	o no-harm principle	o qualified sovereignty
▪ social contract	▪ protection of public health and environment	▪ transboundary impact	▪ equality of states
▪ utilisation of resources for people	▪ notification	▪ state responsibility	▪ liability
o common but differentiated responsibilities	o eco-justice	o common heritage or concern of humankind	o equitable and reasonable utilization
▪ social justice	▪ respect for all life	▪ interdependency	▪ community of interests
▪ fair trade	▪ monitoring	▪ eco-labelling	▪ integrated approach

8.4   Operationalisation of the framework

The operationalisation of the framework *Guardianship over Water* requires further research beyond the scope of this book. Nevertheless, some general remarks can be made on how the framework can be used. The elements of the framework can be balanced on a case by case basis according to their comparative weight and taking the effects at all policy-levels into account. Cooperation between the relevant actors is required in each and every one of the aspects of the framework. Evaluation of its application is a continuing process.

Furthermore, this framework can be further elaborated in the future. Three possible illustrative ways to this end are now suggested. These do not claim in any way to be thoroughly analysed examples, but solely serve to contribute to the instigation off further research. First, a Draft Declaration on Guardianship over Water is presented. Second, a pricing mechanism is suggested which takes into account affordability, cost-recovery and sustainability. Third, a preliminary assessment is made of the application of the framework to evaluate legal instruments at the various policy-levels.

1.   Draft Declaration on Guardianship over Water

The implications of the framework for people, states and the international community as a whole are now summarised in a Draft Declaration on Guardianship over Water. This Draft Declaration articulates sustainable development of freshwater resources as reflected in the framework. The Draft Declaration addresses all people: the people who make up the local community, the state and the international community; the people who both constitute the subjects of international law and decide upon its substance and procedures directly or indirectly. The involvement of and cooperation between all actors is required if the sustainable development of freshwater resources is to be implemented. For people as individuals, groups, states, companies or otherwise to take responsibility, their empowerment is of the utmost importance and interactive capacity-building as well as leadership in service of people and their environment are therefore emphasised.

Guardianship over Water

In order to promote the sustainable development of freshwater resources, international law must construe the role of humankind as that of guardian over water by applying the three principles below. This implies that people, states and the international community as a whole must have access to water to meet their basic needs, and that they have a right to use water for other purposes – such as industrial and cultural uses – as long as the quantity and quality of water required for basic human needs and environmental protection is assured. Access and use of water are required to achieve the development of people, states and the world community. The protection of water at all levels is needed for it to support the life of humans and the environment. Water must be used sustainably by reducing waste and responding to problems of scarcity as well as maintaining ecosystems. The principles are to be viewed in their interrelationship and interdependence.

1. Human rights and duties

1.1   People have a human right to water to meet their basic needs and a right to use water for other purposes, but this right is balanced by their duty and right to protect the environment.

1.2   People have a right to development, which requires both basic access to water and the right to use water for economic purposes, and a duty to respect and protect the aquatic environment.

1.3   People have a right to a healthy environment and a duty to contribute to such an environment.

1.4   People have a duty to take responsibility for their consumption and pollution of water to the end of conserving water and preserving the environment at large.

2. Qualified sovereignty of states

2.1   States have a right to use water in their territory to reduce poverty through the provision of water to their people for basic uses. States also have a sovereign right to use water for economic purposes, qualified by the principle of equitable utilization, the obligation not to cause significant harm in territories beyond their jurisdiction and a duty to protect the environment.

2.2   States have a right and duty to use water for their development and for the implementation of the right of self-determination of peoples, which requires both basic access to water and the right to use water for economic purposes, and a duty to respect and protect the aquatic environment.

2.3   States have a right and duty to use water for the protection of public health and to protect the environment entailing its life support systems.

2.4   States shall ensure that their use of water is not to the detriment of the global waters or the environment at large.

3.   *Equitable and reasonable utilization*

3.1   In the application of the principle of equitable and reasonable utilization, the international community – including people, states, companies and organizations – has a responsibility to safeguard the basic access to water for present and future generations and to stimulate a supportive and open international economic system, qualified by the protection of ecological integrity and the environment as the common concern of humankind.

3.2   The international community shall apply the principle of common but differentiated responsibilities in safeguarding equitable access to and use of water for all people while at the same time protecting the aquatic environment.

3.3   The international community has a responsibility to promote the use of water in line with eco-justice, responding to the need of social equity as well as protection of the environment.

3.4   The international community shall perceive the hydrological cycle as a common heritage of humankind, in order to promote the sustainable use of water and the preservation of the environment at large and safeguard equitable and reasonable utilization of the world's waters.

2.   Three-tranches pricing mechanism

The pricing of water can be used as an economic incentive by which to influence people's behaviour and can therefore be a suitable instrument of water management.[11] In order to stimulate the achievement of sustainable development, a water pricing mechanism needs to take into account such factors as basic human needs and environmental principles, besides economic rationales such as cost recovery.[12] According to Savenije and Van der Zaag: 'in water pricing, adequate attention should be given to equity considerations through, for example, increasing block tariffs.'[13] Block-pricing implies cross-subsidies, enabling both equity and financial sustainability.[14] A block-pricing system dividing the price of water into three tranches that embrace all categories of sustainable development and that could be applied at the national level is now suggested. This three-tranches pricing mechanism is further elaborated upon and represented by a model and accompanying table in Annex II.

---

[11] See on water system rate structure, Table 16, and on water prices for various households, Table 17, of Gleick, Burns, Chalecki, Cohen, Cushing, Mann, Reyes, Wolff and Wong (2002).

[12] This Paragraph and Annex II are based upon Hildering (forthcoming). For a pricing system comparable to the one presented here, see *e.g.* Petrella (2001), pp. 95-97.

[13] Savenije and Van der Zaag (2002), p. 98, continuing: 'Instead of economic pricing there is a need for defining a reasonable price, which provides full cost recovery but which safeguards ecological requirements and access to safe water for the poor.'

[14] Savenije and Van der Zaag (2002), p. 104.

This recommendation is only intended to offer an outline of the factors to be taken into account from an international law perspective and does not reflect an economic analysis as such. Further identification is needed of, *e.g.*, the quantities, the quality required and price of water in the specific circumstances. The amount of water needed obviously differs depending on such local aspects as climate. Moreover, it would be difficult to measure or provide the quantities required in cases where people do not possess land with wells or are not connected to taps or meters.[15]

The three-tranches pricing mechanism reflects the principles of the framework *Guardianship over Water*. The first tranche guarantees the basic amount and quality of water needed; the second one is priced according to the market approach; and the third one discourages unsustainable use.

From the social angle, all people must have basic access to water, even if this means that cost recovery is not always possible.[16] The inequality of incomes needs to be reflected in the price of water.[17] Furthermore, it is often the case that poor people who do not have access to tap water need to buy water in bottles and therefore have to pay more.[18] In identifying the price of water, the polluter and user pays principle should be applied whilst at the same time protecting the poor from unaffordable prices.[19]

From the economic perspective, low prices for water can be said to encourage wasteful use of water resources.[20] In order to be efficient in an economic sense, prices should at least cover the costs.[21] Full-cost recovery

---

[15] In the case where even the basic amount of a certain quality of fresh water is scarce, the problem enters another dimension in which the obligation of states to cooperate will be of great importance and could lead to exchange of (virtual) water.

[16] See, *e.g.*, Caponera (1992), p. 155, who argues that water has a price and that the costs entailed by its development and conservation need to be reimbursed as far as possible by the users, taking into account: 'market forces, social needs, political requirements, public interest, availability of water and, last but not least, the ability of the users to pay.'

[17] Caponera (1992), p. 9, argues that a system of allocating water in accordance with its maximum cost benefit would only be adequate if all users had the same monetary capacity.

[18] For example, in Mexico City poor people in certain areas, not connected to taps, have to buy expensive water while rich people are connected to taps and therefore provided with cheap water. See also Barlow and Clarke (2002), p. 59, for examples in Peru and Honduras.

[19] See, *e.g.*, Global Water Partnership (2000), pp. 36-38. On p. 38 of the Framework for Action, criteria on tariffs are identified.

[20] In both developing and developed countries there are examples of water pricing that are not economically viable. For example, people in Amsterdam and parts of Rotterdam pay a flat fee for their use of water no matter how much they actually use. See Dalhuisen, De Groot and Nijkamp (2000), p. 12. Moreover, in many industrialised countries agricultural use of water is highly subsidised.

[21] See, *e.g.*, World Commission for Water in the 21st Century (2000), p. 63, where it is stated that adoption of full-cost pricing of water use and services is the most important policy recommendation they make, and the EC Directive establishing a framework for Community action in the field of water policy (Water Framework Directive)

is often viewed as the economically sound way of pricing water.[22] When calculating the price of water, its use costs (incurred in financing and operating the abstraction, transmission, treatment and distribution systems) and opportunity costs (imposed on others as a result of use of the water) need to be taken into account.[23] When full-cost recovery is referred to, it often includes the profits required by the private sector. The private sector will be stimulated to invest in the supply of water when the price reflects the full costs.

From a sustainable and environmental viewpoint, issues such as intergenerational equity and the exhaustibility of groundwater must also be taken into account. Water use beyond sustainability is to be discouraged by setting a high price that besides full cost-recovery includes costs for future generations, reflects over-pumping and pollution, and encourages the efficient use of water and investments in water-saving technologies. The parties and uses involved can make such use of water still very lucrative and effective, requiring differentiation. For example, a company producing highly profitable luxury goods but degrading the environment by its use of water will not be discouraged by the same prices that inhibit a consumer using water for washing his or her car every day or keeping a lawn green in an arid region.

Before considering the application of economic incentives such as pricing it must be considered whether waste or mismanagement of water actually takes place and if the incentive is appropriate, efficient and effective. For example, pricing water could in many cultures, especially indigenous ones, have an antagonistic effect, estranging peoples from the water and ground, which is likely to undermine the very goal aimed for: sustainable development.

3.   Possible application to legal instruments

The framework *Guardianship over Water* could furthermore be used to develop, systematically evaluate and assess legal instruments regulating water management and as guidance for new policy-making in view of sustainable development. The result of an evaluation could, for example, be presented in matrices resembling the structure of the comprehensive

---

(2000/60/EC, OJ 2000 L327EC), of which Article 9 requires member states to take account of cost recovery of water services, but this provision can be deviated from.

[22] See, *e.g.*, World Commission for Water in the 21st Century (2000), p. 3, where it is stated that fresh water must be recognized and managed as a scarce commodity and that full-cost pricing of water services will be needed to promote conservation and to attract the required large investments. It continues: 'Polluter pays and user pays principles must be enforced. And mechanisms must be found whereby those who use water inefficiently have incentives to desist and transfer that water to higher-valued uses, including environmental purposes.'

[23] See Briscoe (1996) for an elaboration of the different costs to be reflected in the price of water.

framework. The framework could also be used to identify trade-offs that obstruct sustainable development. A thorough elaboration of such applications takes us beyond the scope of this book. Further examination of the application of the framework to legal instruments could be undertaken in future research, possibly including field research.

As a first step, a preliminary assessment of a legal instrument at each of the three policy-levels – community, national and international – is now provided to illustrate the possible application of the framework. The three examples are taken from instruments that express at least a concern for sustainable development. The illustrative example at the community, national and international level are, respectively: the Madya Pradesh Watershed Management framework, the Constitution of the Republic of South Africa and the EU Water Framework Directive.

Madya Pradesh Watershed Management
At the community level, a preliminary application of the framework to the Madya Pradesh Watershed Management framework, part of the 1994 Rajiv Gandhi Mission on Watershed Management, in India, provides the following notion.[24] The management takes a community approach. Access to water for basic human needs is likely to be encouraged by community involvement. People can practice a right to use water since they manage the water. A balance is drawn up with their duty to protect the water, addressing the needs of the environment. The combination of development and environment is likely to result in a healthier environment. In short, human rights and duties appear to have been balanced here. The well-being of people can be expected to thrive most from this approach, in addition to ecological gain and economic prosperity. Management of the watershed in line with sustainable development is likely to have a positive impact at both the national and international level. The Mission on Watershed Management involves all interested groups, especially at the local level, and pays special attention to groups such as women. Landowners seem to profit most from the successes of the Mission. Increased involvement is combined with an ecological approach, addressing poverty and environment at the same time, and also increasing the amount of usable land for agriculture. The Mission seems to show that, although further work has to be accomplished, sustainable development can be implemented in such a way as to favour all pillars. Whether the polluter and user pays principle is applied remains unclear. Moreover, the inequality in India, in this case specifically between landowners and non-landowners, requires further attention, as the evaluation report of the Mission also makes clear.

---

[24] See, *e.g.*, Rajiv Gandhi National Drinking Water Mission (1993), *An introduction to Rural Water Supply and Sanitation Programmes in India*, Ministry of Rural Development: New Delhi.

Constitution of the Republic of South Africa
At the national level, a preliminary application of the framework to the Constitution of the Republic of South Africa is now undertaken.[25] The 1996 new and progressive Constitution of South Africa includes a Bill of Rights in Chapter 2.[26] In this Bill of Rights, people are granted a right of access to health care, food, water and social security. According to the Bill of Rights: 'Everyone has the right to have access to sufficient food and water.'[27] The environment is to be protected for present and future generations. In the process of drafting the Constitution, a public participation programme was conducted. In addition, for example, equal access to administration is provided for. The Constitution promises stability by overcoming the divisions of the past, and could therefore provide for a healthy economic environment as well. The Constitution is likely to promote the eradication of poverty, provide for equitable use, and protect the environment. Although it is a national instrument, its impact at the local level might be even more positive. At the international level, it could have a positive impact by providing an example for other countries in the region (SADC) as well as globally, in integrating concerns relating to sustainable development into a constitution or other documents. The main concern seems to be the actual implementation of the Constitution considering, among other things, the required finances.[28]

EU Water Framework Directive
At the international – or supranational – level, a preliminary application of the framework to the EU Water Framework Directive (EUWFD) provides the following overview.[29] The EUWFD integrates and supplements earlier EU directives on various water uses such as drinking and bathing, enhancing predictability, clarity and determinacy through an integrated approach. The EUWFD is intended to promote sustainable, balanced and equitable water use. The Directive contains social, economic and ecological provisions such as on participation, pricing and a "good ecological status" of a body of surface water. Article 2.21 defines ecological status as 'an expression of the quality of the structure and functioning of aquatic ecosystems associated with surface waters, classified in accordance with Annex V'.

---

[25] Constitution of the Republic of South Africa 1996, Act 108 of 1996, as adopted on 8 May 1996 and amended on 11 October 1996. Entry into force: 7 February 1997.
[26] In the 1994 Bill of Rights of the new Constitution of South Africa, several social, economic and environmental rights are protected. See on environmental rights in the South African Constitution, *e.g.*, Glazewski (1996).
[27] Section 27(1)(b) of the Bill of Rights.
[28] According to Gleick (2000), p. 9, water policies are being developed to implement the right to water.
[29] Directive 2000/60/EC of the European Parliament and of the Council of 23 October 2000 establishing a framework for Community action in the field of water policy, *Official Journal of the European Communities*, 22 December 2000, L 327. See on sustainability and the EUWFD, *e.g.*, Rieu-Clarke (2004).

Integrated water management is based upon river basin districts. The river basin approach is likely to have a positive effect, also on states beyond EU borders: member states are encouraged to establish coordination with non-Member states to which the River Basin District extends 'with the aim of achieving the objectives of this Directive throughout the River Basin District' (Article 3.5). Member States are also to encourage involvement of all interested parties and make certain information available for comments. The emphasis on participation is likely to lead to a comparative improvement in the opportunities for people to exert influence. In Article 13.4 of the EUWFD, it is stated that river basin management plans are to include information detailed in Annex VII. In this Annex VII, on river basin management plans, the elements to be covered include 'a summary of the public information and consultation measures taken, their results and the changes to the plan made as a consequence'. Article 14 of the EUWFD deals with public information and consultation, *e.g.* stating that: 'Member States shall encourage the active involvement of all interested parties in the implementation of this Directive, in particular in the production, review and updating of the river basin management plans.' Continuous water resource assessment is also included. Cost recovery is recommended but is not formulated as a legal obligation. Many of the elements of sustainable development appear to be present in the Directive. Whether, for example, the possible derogations from "good ecological status" for social and economic reasons in practice result in a balanced approach to sustainable development or in trade-offs presenting a bias needs to be further reviewed. Moreover, the time-schedule for implementation might be unrealistic in some aspects. Also in view of possible effects beyond EU borders, the developmental part might require closer attention.

8.5   Conclusions

Building on the outcomes of the foregoing chapters that systematically analysed the different elements of sustainable development in water management from an international law perspective, this Chapter balanced all three pillars of sustainable development. In combining the social, economic and ecological categories, the following can be concluded.

Sustainable development of water integrates human rights and duties at the community level, qualified sovereignty of states at the national level, and a modified principle of equitable and reasonable utilization of water at the international level. The human rights and duties of people include a human right to water, the right to use water, a duty to protect water, a right to development, a right to a healthy environment and a duty to pay for pollution or consumption. The sovereignty of states is qualified by rights and duties that result from eradication of poverty, water as an economic good, environmental protection, the right of self-determination of peoples, the precautionary principle and the no-harm principle. Equitable and reasonble utilization of water is to reflect equity, a supportive and open international economic system, ecological integrity, common but differentiated responsibilities, ecological justice, and a common concern or heritage of humankind.

All three pillars of sustainable development, the policy-levels, the key concepts and the principles of international law combined can be articulated by *guardianship over water*. *Guardianship over water* integrates *access to water* within the social pillar, *control over water* within the economic pillar, *protection of water* within the ecological pillar, *development through water* combining the social and economic pillars, *life support by water* combining the social and ecological pillars, *sustainable use of water* combining the economic and ecological pillars and *sustainable development of water* combining all three pillars.

The comprehensive framework *Guardianship over Water* identifies the principles of international law that can jointly contribute to sustainable development in and through water management and clarifies their relationship. Therefore, the framework answers the research questions. It can, for example, be used to communicate sustainable development of water to state and non-state actors through a declaration, to provide guidance in the use of water pricing mechanisms and to evaluate legal instruments regulating water management.

The overall assessment of the research and the resulting conclusions and recommendations are formulated in the following concluding Chapter.

Conclusions

Water is a condition necessary for life. The world's surface largely consists of water. Still, that same world faces a water crisis. Only a small part of the global waters is fresh water, most of which is hidden in ice caps and deep underground. The water resources accessible for humans are dispersed over the earth in an unequal fashion and often polluted. More than 1 billion people do not have access to drinking water and over 2 billion people lack access to basic sanitation. As a consequence, people are dying and struck by illness. Nature is suffering as well because of too much, too little or degraded water. Besides the social and ecological considerations, the third pillar of sustainable development – the economy – cannot do without water either: almost any production process requires water. The water crisis is mainly viewed as caused by governance failure and the lack of people's awareness of the gravity of the problems.

In light of the above, this book addresses the following questions:

- *Which principles of international law can be instrumental in achieving sustainable development in water management?;*
- *How do they relate to one another?; and*
- *How can they jointly contribute to a more sustainable development of freshwater resources?*

To address the research questions, the different elements of international law for sustainable development in water management have been systematically analysed in this book. The methodology adopted in this study can be divided into four steps. The first step is the identification of the present state of sustainable development concepts in international water law. The second step is the development of a comprehensive framework of principles of international law. As a third step, this book has been constructed on the basis of the new structure. The methodology and the developed framework are therefore reflected in the structure of the book: Part I introduces and defines the terms of reference in addressing the uses of freshwater resources (Chapter 2), and analyses the way sustainable development is embedded in international law (Chapter 3). In Part II, an analysis is undertaken within each of the sustainable development pillars to identify the demands on international law made by water as respectively a social (Chapter 4), economic (Chapter 5), and ecological (Chapter 6) good. The combination of principles of international law relevant to sustainable development of freshwater resources is the subject of Part III. Chapter 7 bridges the gap between the social and economic, the social and ecological and the economic and ecological pillars of sustainable development. In Chapter 8 all three pillars of sustainable development are combined, resulting in the comprehensive framework *Guardianship over Water*. The fourth step, an overall assessment of how international law can contribute to the achievement of sustainable development in and through water management, is now undertaken.

The analysis per chapter has resulted in seven groups of inferences:

1. The analysis in Chapter 2 inferred that: i) The potential conflicts between uses and users of water are numerous and severe. ii) The allocation of fresh water between its various uses has become a major issue at the community, national as well as the international level. iii) Water problems call for a balance of interests that requires international law to take an integrated approach.

2. The analysis in Chapter 3 inferred that: i) International water law has evolved in a fragmented manner. ii) The concept and objective of sustainable development has gained much support reflected, *e.g.*, in an emerging international law on sustainable development that builds on human rights law, international development law and international environmental law. iii) The achievement of sustainable development requires adjustment of the principle of equitable and reasonable utilization of water resources and increased reflection in international law of the trends toward integration, cooperation and community interests.

3. The analysis in Chapter 4 inferred that the key concept of water as a social good, *access to water*, includes a human right to water at the community level, eradication of poverty at the national level, and the principle of equity at the international level. The Chapter inferred that: i) A human right to water should be further affirmed by states. ii) International law does not yet sufficiently reflect the necessity and reality of non-state actor participation. iii) The realisation of access to water calls for further eradication of poverty and application of equity.

4. The analysis in Chapter 5 inferred that the key concept of water as an economic good, *control over water*, includes a right to use water at the community level, water as an economic good at the national level, and a supportive and open international economic system at the international level. The Chapter inferred that: i) Ownership of water in principle concerns user rights that are preferably regulated and controlled by democratic public bodies. ii) Community-public-private partnerships can under conditions provide a promising way to manage water. iii) An economic approach to water can assist in the efficient management of water but is not necessarily compatible with sustainable development.

5. The analysis in Chapter 6 inferred that the key concept of water as an ecological good, *protection of water*, includes a duty to protect water at the community level, protection of the environment at the national level, and ecological integrity at the international level. The Chapter inferred that: i) Water as an ecological good requires better protection by international law, including enhancement of the catchment area approach. ii) The process of equitable and reasonable utilization is inconclusive on the protection offered to ecosystems and water. iii) Preventive protection of water is to be preferred over measures to combat degradation of water.

6. In Chapter 7 the social and economic, social and ecological, and economic and ecological pillars are combined, respectively resulting in the following inferences: i) *Development through water* includes the right to development, the right of self-determination and the principle of common but differentiated responsibilities. ii) *Life support by water* combines social and ecological interests through the right to a healthy environment, the precautionary principle and eco-justice. iii) *Sustainable use of water* consists of the polluter and user pays principle, the no-harm principle and the common heritage or concern of humankind.

7. In combining all three pillars of sustainable development, the research in Chapter 8 resulted in the following inferences: i) *Sustainable development of water* integrates human rights and duties at the community level, qualified sovereignty of states at the national level, and a modified principle of equitable and reasonable utilization of water at the international level. ii) All three pillars, the policy-levels, the key concepts and the principles of international law combined can be articulated by *guardianship over water*. iii) The comprehensive framework *Guardianship over Water* identifies the principles of international law that can jointly contribute to sustainable development in and through water management and, therefore, answers the research questions.

Together the inferences result in three key conclusions:
1. The current international law on freshwater resources does not necessarily contribute to sustainable development and although international law on sustainable development is emerging, its application to freshwater resources remains unclear.
2. Numerous water laws and principles of sustainable development exist at the community, national and international level but there is no textual formulation in international law of sustainable development of water that addresses all parties, includes all policy-levels, and converges the several chapters of international law.
3. International law is to make sure that water is not only dealt with as an economic good, but also as a social and ecological good.

Three key recommendations are now made in response to those conclusions:
1. The comprehensive framework *Guardian over Water* identifies the principles of international law that can jointly contribute to sustainable development in water management. It thereby provides increased transparency to the interrelationship between those principles, revealing the potential for balancing social, economic and ecological interests. The framework could, for example, be used by lawyers and policy-makers to assess the adequacy of international law instruments and of water management in achieving sustainable development.

2. The Draft Declaration on Guardianship over Water provides for a means to communicate sustainable development of freshwater resources to all people at all policy-levels.
3. When pricing water, a three-tranches pricing mechanism – an application of block-pricing – is recommended to reflect water as a social, economic and ecological good.

Annex I.   Principles per policy-level

The community level: human rights and duties

At the community level the following principles can be identified on the basis of the foregoing chapters.

The principle within the social pillar at the community level is a human right of people to water (4.2). The right to water is, moreover, a necessary condition for the fulfilment of other human rights. The underlying reason for the right to water is that water is a basic human need. In order to realise the human right to water, a reporting procedure (see also 3.2.1) is required.

Human Right to Water	Reason	Tool
Community level	basic human need	reporting

The principle within the economic pillar at the community level is the right of people to use water (5.2). This principle is a pre-condition for development. The underlying reason for the right to use water is that everybody has a right to a certain independence enabling individual freedom. In order to realise the right to use water, community management (5.3.2) is recommended, with special regard for the position of vulnerable groups (4.3.2).

Right to Use Water	Reason	Tool
Community level	individual freedom	community management

The principle within the ecological pillar at the community level is the duty of people to protect water (6.2). The underlying reason for the duty to protect water is that all people share in the collective responsibility for their environment. The realisation of the duty to protect water calls for an increased awareness of and involvement of people in dealing with water issues (4.2.3).

Duty to Protect Water	Reason	Tool
Community level	collective responsibility	community involvement

The principle within both social and economic pillars at the community level is the right of people to development (7.2.1). Although this principle is a condition for the fulfilment of other human rights, its status as a separate human right is controversial. The underlying reason for the right to development is that people must have the opportunity to reach a certain quality of life. Pure survival does not do justice to the human dignity underlying human rights. In order to realise the right to development, the key is that education (4.3.1) must be accessible for all and with special regard for vulnerable groups (4.3.2).

Right to Development	Reason	Tool
Community level	quality of human life	education

The principle within both social and ecological pillars at the community level is the right of people to a healthy environment (7.3.1). Whether or not this principle is an emerging human right is controversial. The underlying reason for the right to a healthy environment is that even a minimal quality of life demands a certain quality of living environment. In order to come to the realisation of the right to a healthy environment, price differentiation (8.4 and Annex II) can be used to safeguard basic water needs and discourage unsustainable consumption behaviour.

Right to a Healthy Environment	Reason	Tool
Community level	quality of living area	price differentiation

The principle within both economic and ecological pillars at the community level is the polluter and user pays principle (7.4.1). This principle directly links the costs to the use of water and thereby provides for an economic incentive for sustainable use of water. The underlying reason for the polluter or user pays principle is that people must take responsibility for their actions. In order to realise the polluter pays principle, such side effects of consumption as environmental pollution must be reflected in the pricing of goods and services, internalising such externalities (Annex II) as far as possible.

Polluter and User Pays Principle	Reason	Tool
Community level	individual responsibility	internalisation of externalities

The principle integrating all pillars at the community level is that of human rights and duties (see also 3.2.1). The reason underlying human rights is that of human dignity. Moreover, respect and a sense of pride in oneself go hand in hand with taking responsibilities. Human rights and duties reflect both the general rule that rights cannot be defined without duties and the requirement for a balanced approach. In order to realise human rights and duties, subsidiarity (1.4) is best applied.

Human Rights and Duties	Reason	Tool
Community level	human dignity	subsidiarity

The national level: qualified sovereignty

On the basis of foregoing chapters, the following principles can be identified at the national level.

The principle within the social pillar at the national level is the duty of states to eradicate poverty (4.3). This principle appears to be well-established, at least as a political and moral obligation of the state. The underlying reason for the duty of states to eradicate poverty is that people have entrusted their government with the task of taking care of collective interests, requiring solidarity. In order to achieve such eradication of poverty, the significance of capacity-building (4.3.1) cannot be overestimated.

Eradication of Poverty	Reason	Tool
National level	solidarity	capacity-building

The principle within the economic pillar at the national level is that of water as an economic good (5). This principle seems to be emerging in international law. The underlying reason for regarding water as an economic good is that of economic viability. In order to realise the functioning of water as an economic good, pricing can be an important instrument (8.4 and Annex II).

Water as an Economic Good	Reason	Tool
National level	economic viability	pricing

The principle within the ecological pillar at the national level is the duty of states to protect the environment (6.3). This principle appears to be rapidly emerging as an established principle of international law. The underlying reason for the duty of states to protect the environment is the wide-spread degradation of water systems and other ecosytems. An appropriate instrument for the realisation of such protection is provided by the environmental impact assessment (6.2.3).

Protection of the Environment	Reason	Tool
National level	degradation of nature	environmental impact assessment

The principle within both social and economic pillars at the national level is the right of self-determination of peoples (7.2.2). This principle appears to be well-established in international law. The underlying reason for the right of self-determination is the philosophy that a state is based on a social contract with its people, a contract that can be annulled when a state looses its legitimacy. In order to fulfil the requirements of the right of self-determination, a state must utilise natural resources to the benefit of their people (7.2.2).

Right of Self-determination	Reason	Tool
National level	social contract	utilisation of resources for people

The principle within both social and ecological pillars at the national level is the precautionary principle (7.3.2). The status of this principle is that of an emerging principle of international law. The reason underlying the duty of a state to take a precautionary approach is that of its responsibility for the well-being of its people. In order to realise such protection, states must adhere to, for example, notification (3.4.2).

Precautionary Principle	Reason	Tool
National level	protection of public health and environment	notification

The principle within both economic and ecological pillars at the national level is the no-harm principle (7.4.2). This principle is well-established within international law and can be said to be part of customary international law as well as treaty law. The reason underlying the duty of states to take all appropriate measures to prevent or otherwise combat significant harm caused by them in territory beyond their jurisdiction, is that states are liable for effects of their actions beyond their jurisdiction. In order to realise the principle, state responsibility (7.4.2) is to be applied.

No-harm Principle	Reason	Tool
National level	transboundary impact	state responsibility

The principle integrating all pillars at the national level is that of the qualified sovereignty of states (see also 5.2.2). This principle is well-established, especially within international water law. The underlying reason for the right of states to practice their sovereignty within limitations is the equality of states together with the fact that use of water resources by one state will practically always have an impact on the use of water by another state. In order to practice sovereignty over freshwater resources within its qualifications, states need to be able to be held liable for their behaviour and uses (7.4.2).

Qualified Sovereignty	Reason	Tool
National level	equality of states	liability

The international level: equitable and reasonable utilization

On the basis of the foregoing chapters, the following principles can be identified at the international level.

The principle within the social pillar at the international level is that of equity, both intra- and intergenerational (4.4). Equity can be argued to be a general principle of law, but both intra- and intergenerational equity appear to be emerging principles of international law. As concepts, intra- and intergenerational equity are frequently used in international politics. The underlying reason for the equity principle lies in the need for a certain degree of fairness in the relations among the international community. In order to realise such equity, the present generation must be given access to information and justice (3.2 and 4.2.3), while future generations must also be allowed to be represented.

Equity	Reason	Tool
International level	fairness	access to information and to justice

The principle within the economic pillar at the international level is that of a supportive and open international economic system (5.4). This principle can be argued to be emerging within international law as reflected in, for example, Principle 12 of the Rio Declaration and WTO regulation. The underlying reason for the responsibility of the international community to come to such an international economic system is that in the long-term trade and investment cannot be effective if they do not accord with such a system. In order to come to the realisation of such a system, water must not be viewed a commodity in itself but can, for example, be more sustainably used through common management (3.4.3).

Supportive and Open International Economic System	Reason	Tool
International level	long-term effectiveness	common management

The principle within the ecological pillar at the international level is that of ecological integrity (6.4). Although elements of ecological integrity such as the protection of the marine environment are well-established, as a principle of international law ecological integrity can at the most be argued to be progressively developing. The underlying reason for the responsibility of the international community to safeguard the ecological integrity is based on its interests in the hydrological cycle. In order to act upon such responsibility and interests, a catchment basin approach (6.4.1) must be taken in the management of freshwater resources.

Ecological Integrity	Reason	Tool
International level	interdependency	catchment basin approach

The principle within both social and economic pillars at the international level is that of common but differentiated responsibilities (7.2.3). This principle can be said to be well-established within international law even if not actually resulting in the actions required by it. The underlying reason for the international community to assume common but differentiated responsibilities, is that of social justice, requiring the assumption of responsibilities according to the contribution to the problem and also emphasising solidarity. The realisation of such responsibilities requires, *e.g.*, fair trade (5.5.).

Common but Differentiated Responsibilities	Reason	Tool
International level	social justice	fair trade

The principle within both social and ecological pillars at the international level is eco-justice (7.3.3). This principle cannot be said to be established within international law but those elements derived from human rights are, variously, either well-established or emergent. The basic reason for the responsibility of the international community to provide for eco-justice lies in the conviction that both human and other forms of life should be respected, which increasingly is also becoming a necessity. In order to realise eco-justice, knowledge of both protection and violations of human rights and ecological interests needs to be available, in the first place requiring monitoring (7.3.3).

Eco-justice	Reason	Tool
International level	respect for all life	monitoring

The principle within both economic and ecological pillars at the international level is that of the common heritage or concern of humankind (7.4.3). The principle of common heritage of humankind cannot be said to be well-established within international law beyond the realm of the law of the sea and outer space and its further emergence remains uncertain; although it is part of a limited number of treaties and other legal instruments, in the more recent documents of international law it is replaced by the principle of common concern of humankind. The underlying reason for a common heritage or concern of humankind is the interdependency between humans in both time and space. In order to realise such care-taking, ways to contribute to it need to be evident, for example, by eco-labelling (7.4.3).

Common Heritage or Concern of Humankind	Reason	Tool
International level	interdependency	eco-labelling

The principle integrating all pillars at the international level is the principle of equitable and reasonable utilization (see also 3.3.2) under the condition that it aims for the achievement of sustainable development. The underlying reason for such utilization of water resources is the community of interests shared by the international community such as in human well-being and welfare and in the environment. An integrated approach (3.4.1), *e.g.* taking into account all actors and policy-levels, is indispensable.

Equitable and Reasonable Utilization	Reason	Tool
International level	community of interests	integrated approach

Annex II. The price of water

The actual effect of a change in price, the elasticity of water, is debated. The elasticity may vary depending on economic factors such as the Gross Domestic Product (GDP) per capita of a country: at higher GDP levels, pricing of water may have more influence on consumer behaviour.[1] According to Dalhuisen: 'Increasing block rates often result in a more elastic price elasticity of water demand.'[2]

A complication in pricing water is posed by the many externalities involved.[3] Internalisation of externalities requires the valuation, for example, of human life and health and biodiversity. In developing countries most diseases are water related, but the price paid by a country in loss of productivity is not usually reflected in the price and value of water.[4] For a pricing mechanism to work and potentially contribute to sustainable development, externalities will have to be internalised to a much greater extent. For example, public regulation could influence the price to reflect the pollution associated with the water use. At the international level this would require cooperation by means of, for example, joint institutions.[5] Price wars by large companies in order to strengthen their position may also lead to imperfect pricing that does not reflect the value of goods.[6]

The valuation of social and ecological impacts is particularly subject to insights that change over time, emphasising the need to frequently recon-

---

[1] See Dalhuisen (2002), p. 145.

[2] Dalhuisen (2002), p. 145. On p. 15, he describes 'increasing block-rate pricing' as a price that is constant within discrete intervals of use, but increasing between different intervals.

[3] On externalities in the context of water pollution, see Murty, James and Misra (1999), pp. 6-8. On p. 6 it is explained that: 'An externality is present whenever individual A's utility and production relationships include real (*i.e.* non-monetary) variables, whose values are chosen by others (persons, corporations, governments) without particular attention to the effects on A's welfare (Baumol and Oates, 1988).' Externalities can be both positive as well as negative, such as environmental pollution. In order to create information on the value of water in certain circumstances, encompassing all elements of the available economic options, including valuation of preservation, it is best if the externalities are internalised. However, it is almost impossible to transfer the various values of water into entities such as numbers or money necessary for comparing values as well as for internalising externalities.

[4] Another example is the labour of women, for example domestic labour and agriculture, which is often not paid and therefore not translated into economic values.

[5] See Murray, James and Misra (1999), p. 30: 'While economic instruments can be used by national governments to deal with national or local externalities, alternative institutions are needed to support economic instruments to deal with transnational or global externalities.'

[6] Large companies can very well be in the position to play the market. By lowering its price beyond cost recovery a large company can cause competition to disappear and create a monopoly.

sider the valuation. Although there is a distinction between the valuation and pricing of water needs, they are strongly related: 'Water's *value* is important in deciding on alternative uses of the scarce resource whilst *charging* is an economic instrument to recover costs and provide incentives for efficiency and conservation.'[7] Valuing water economically can pose some problems, as in the case where basic domestic needs are valued lower than industrial use, requiring intervention by means of national or international public law.[8] Another difficulty is that, despite much research, there is not enough information available on such factors as the flow and impact of water on economy, people and environment. Even if such information is available, it is difficult to value changes in water quality and its implications for, *e.g.*, human health and biodiversity.[9]

With the above in mind, the three-tranches pricing mechanism recommended in Section 8.4 is now further elaborated upon and represented in a model and accompanying table. In the model following below, p stands for price and q for quantity of fresh water. Tranche 1 (social) consists of a single price in order to avoid a complicated and expensive system. This tranche is mainly a social tranche, reflecting a social perspective in considering the basic need of people, and relates to the domestic uses of fresh water. A (human) right of access to this water could be implemented through this tranche. The price can range from subsidised to a market price, depending on the ability of people in a society to pay for water. If the price of water causes basic products such as food to become too expensive for some population groups, measures need to be taken to support these groups. There may also be a need for (temporary income) subsidies to overcome problems such as encountered with the low food grain prices, complicating full cost-recovery of irrigation.[10] Despite the attraction of subsidies, it is advisable only to use that instrument as a means of last resort, and temporarily, since practice has shown that it often promotes inefficiency. Tranche 2 (economic) consists of a market price. The price should not be too differentiated to avoid a complicated as well as expensive sys-

---

[7] Global Water Partnership (2000), p. 35. As stated in the Framework for Action, on the same page, realigning economic and financial practices is at the heart of the World Water Vision.

[8] For the need and potential for government intervention in the market, see Dalhuisen, De Groot and Nijkamp (2000), pp. 12-14, who identify the following means of intervention by the government: 'pricing policies (taxes and subsidies), regulation (imposing standards and norms), technical intervention (*i.e.* stimulating the development and adoption of new technologies), the provision of information aimed at increasing awareness of the scarcity of water, and deregulation.'

[9] See Magat, Huber and Viscusi (2000), p. 1: 'While water is undeniably one of our most fundamental and highly valued natural resources, it has been difficult to assign a value to improvements in water quality.'

[10] See Bhatia (2000), p. 3. See also Postel (1996), p. 55: 'There is a broad spectrum of options between full-cost pricing, which could put many farmers out of business, and a marginal cost of nearly zero to the farmer, which is a clear invitation to waste water.'

tem. Here the market approach and economic efficiency come into full effect. In tranche 3 (environmental) the price is much more flexible and depends on the ability of the user to pay in order to discourage this use of fresh water.

The prices of and boundaries between these tranches will vary according to local circumstances, such as availability and quality of water, incomes, transport costs, and the degree of unsustainability. Depending on such circumstances, within a tranche the prices may also vary. It is not advisable to complicate the (administrative) system too much since that would most likely increase the costs of the system, possibly even outweighing the benefits, and could lead to a decrease in transparency and therefore in efficiency. Tranche 1 must be guaranteed under all circumstances. Tranches 2 and 3 relate to all uses except for the basic domestic uses. Their prices will take into account such costs as environmental costs and will be guided by, for example, the polluter and user pays principle and intergenerational equity.

In the table, price 1 will be set at a level guaranteeing the ability to fulfil average basic needs. Price 2 will be the market price including costs such as for the environment and future generations. Price 3, in addition to the economic effective price, will depend on the amount needed to discourage such use. The quantity of water falling within Price 2 and 3 depends on the available amount of water and the impact of its use. They are therefore only referred to as a and b.

Consumer		
Price 1	Tranche 1	On average 50 litres per person per day
Price 2	Tranche 2	a
Price 3	Tranche 3	b

Bibliography

Abouali, G. (1998), 'Natural Resources under Occupation: The status of Palestinian water under international law', in *Pace International Law Review*, 10 (1998) 411.

Abramovitz, J.N. (1996), *Imperiled Waters, Impoverished Future: The decline of freshwater ecosystems*, Worldwatch Paper 128.

Anderson, M.R. (1996), 'Individual Rights to Environmental Protection in India', in Boyle and Anderson (eds) (1996), 199-225.

Arts, K. (2000), *Integrating Human Rights into Development Cooperation: The case of the Lomé Convention*, Kluwer Law International: The Hague.

Arts, K. (2003), 'ACP-EU Relations in a New Era: The Cotonou Agreement', in *Common Market Law Review*, 40 (2003), pp. 95-116.

Baim, K.A. (1997), 'Come Hell or High Water: A water regime for the Jordan river basin', in *Washington University Law Quaterly*, 75 (1997) 919.

Banerjee, U.C. (1992), 'Transfer of Technology: A matter of principles', in Chowdhury, Denters and De Waart (eds) (1992), 311-313.

Barlow, M. and T. Clarke (2002), *Blue Gold: The battle against corporate theft of the world's water*, Earthscan: London.

Baumann, C. (2001), 'Water Wars: Canada's upstream battle to ban bulk water export', in *Minnesota Journal of Global Trade*, 10 (2001) 109.

Benvenisti, E. (1996), 'Collective Action in the Utilization of Shared Freshwater: The challenges of international water resources law', in *American Journal of International Law*, 90 (1996) 384.

Benvenisti, E. (2002), *Sharing Transboundary Resources: International law and optimal resource use*, Cambridge University Press: Cambridge.

Berber, F.J. (1959), *Rivers in International Law*, The London Institute of World Affairs: London.

Bergh, J.C.J.M. van den (1996), *Ecological Economics and Sustainable Development: Theory, Methods and Applications*, Edward Elgar: Cheltenham.

Bergh, J.C.J.M. van den (1999), *Handbook of Environmental and Resource Economics*, Edward Elgar: Cheltenham.

Bertels, J., H. Aiking and P. Vellinga (1999), 'Effects of human activities', in Vellinga and Van Drunen (eds) (1999), 117-150.

Bhatia, R. (2000), *Water & Economics*, Report on the session, Second World Water Forum, in World Water Council (2000).

Biermann, F. (1995), *Saving the Atmosphere: International law, developing countries and air pollution*, Peter Lang: Frankfurt am Main.

Birnie, P.W. and A.E. Boyle (2002) (2nd edition), *International Law and the Environment*, Clarendon Press: Oxford.

Biswas, A.K. (1999), *The Water Crisis: Current perceptions and future analysis*, FIDIC 99 Conference, 19-23 September 1999, The Hague, http://fidic.org/conference/1999/talks/monday/biswas.html.

Boadu, F.O. (1998), 'Relational Characteristics of Transboundary Water Treaties: Lesotho's water transfer treaty with the Republic of South Africa', in *Natural Resources Journal*, 38 (1998) 381.

Boisson de Chazournes, L. (1998), 'Elements of a Legal Strategy for Managing International Watercourses: The Aral Sea Basin', in Salman and Boisson de Chazournes (eds) (1998), 65-76.

Bosnjakovic, B. (1998), 'UN/ECE Strategies for Protecting the Environment with Respect to International Watercourses: The Helsinki and Espoo conventions', in Salman and Boisson de Chazournes (eds) (1998), 47-64.

Bourne, C.B. (1996), 'The International Law Association's Contribution to International Water Resources Law', in *Natural Resources Journal*, 36 (1996) 155.

Boyle, A.E. (1996), 'The Role of International Human Rights Law in the Protection of the Environment', in Boyle and Anderson (1996), 43-69.

Boyle, A.E. (1999), 'Codification of International Environmental Law and the International Law Commission: Injurious consequences revisited', in Boyle and Freestone (eds) (1999), 61-85.

Boyle, A.E. and M.R. Anderson (eds) (1996), *Human Rights Approaches to Environmental Protection*, Clarendon Press: Oxford.

Boyle, A.E. and D. Freestone (eds) (1999), *International Law and Sustainable Development: Past achievements and future challenges*, Oxford University Press: Oxford.

Brans, E.H.P., E.J. de Haan, A. Nollkaemper and J. Rinzema (eds) (1997), *The Scarcity of Water: Emerging legal and policy responses*, Kluwer Law International: London.

Briscoe, J. (1996), *Water as an Economic Good: The idea and what it means in practice*, paper presented at the World Congress of the International Commission on Irrigation and Drainage, Cairo, September 1996.

Browder, G. and L. Ortolano (2000), 'The Evolution of an International Water Resources Management Regime in the Mekong River Basin', in *Natural Resources Journal*, 40 (2000) 499.

Brown Weiss, E. (1989), *In Fairness to Future Generations: International law, common patrimony, and intergenerational equity*, The United Nations University: Tokyo.

Brownlie, I. (1998) (5th edition), *Principles of Public International Law*, Clarendon Press: Oxford.

Bruhács, J. (1993), *The Law of Non-navigational Uses of International Watercourses*, Martinus Nijhoff: Dordrecht, translation of *Nemzetközi vízjog*, translated by M. Zehery.

209

Brunnée, J. and S.J. Toope (1997), 'Environmental Security and Freshwater Resources: Ecosystem regime building', in *American Journal of International Law*, 91 (1997), 26-59.

Bulajić, M. (1988), 'Principles of International Development Law: The right to development as an inalienable human right', in De Waart, Peters and Denters (1998), 359-369.

Bulajić, M. (1993) (2nd edition), *Principles of International Development Law: Progressive development of the principles of international law relating to the New International Economic Order*, Martinus Nijhoff: Dordrecht.

Caflish, L. (1992), 'Règles Générales du Droit des Cours d'Eau Internationaux, Recueil des Cours, 219 (1989-VII), p. 9.

Caflish, L. (1998), 'Regulation of the Uses of International Watercourses', in Salman and Boisson de Chazournes (eds) (1998), 3-16.

Campins-Eritja, M. and J. Gupta (2002), 'Non-State Actors and Sustainability Labelling Schemes: Implications for international law', in *Non-State Actors and International Law*, 2 (2002) 3, 213-240.

Caponera, D.A. (1992), *Principles of Water Law and Administration: National and international*, A.A. Balkema: Rotterdam.

Cassese, A. (2001), *International Law*, Oxford University Press: Oxford.

Chatterjee, S.K. (1992a), 'The International Monetary Fund', in Fox (ed.) (1992), 81-116.

Chatterjee, S.K. (1992b), 'The World Bank', in Fox (ed.) (1992), 119-145.

Chave, P.A. (2001), *The EU Water Framework Directive: An introduction*, IWA Publishing: London.

Chowdhury, S.R., H.M.G. Denters and P.J.I.M. de Waart (eds) (1992), *The Right to Development in International Law*, Martinus Nijhoff: Dordrecht.

Churchill, R.R. and A.V. Lowe (1999) (3rd edition) , *The Law of the Sea*, Manchester University Press: Manchester.

Crawford, J. (ed.) (1988), *The Rights of Peoples*, Clarendon Press: Oxford.

Dalhuisen, J.M. (2002), *The Economics of Sustainable Water Use: Comparisons and lessons from urban areas*, Tinbergen Institute Research Series no. 290, Thela Thesis: Amsterdam.

Dalhuisen, J.M., H.L.F. de Groot and P. Nijkamp (2000), 'The Economics of Water: A survey of issues', in *International Journal of Development Planning Literature*, 15 (2000) 1, 3-20.

Davey, W.J. (1995), 'The WTO/GATT World Trading System: An overview', in Pescatore, Davey and Lowenfeld (eds) (1995), 1-86.

Davis, R.J. (1991), 'Atmospheric Water Resources Development and International Law', in *Natural Resources Journal*, 31 (1991) 11.

Dellapenna, J.W. (2001), 'Foreword: Bringing the customary international law of transboundary waters into the era of ecology', in *International Journal of Global Environmental Issues*, 1 (2001) 3/4, 243-249.

Disanayaka, J.B. (2000), *Water Heritage of Sri Lanka*, University of Colombo: Sri Lanka.

210

Dunoff, J.L. (1994), 'Resolving Trade-Environment Conflicts: The case for trading institutions', in *Cornell International Law Journal*, 27 (1994) 607.

Eckstein, G. (1995), 'Application of International Water Law to Transboundary Groundwater Resources, and the Slovak-Hungarian Dispute over Gabčikovo-Nagymaros', in *Suffolk Transnational Law Review*, 19 (1995) 67.

Eckstein, G. and Y. Eckstein (2003), 'A Hydrogeological Approach to Transboundary Ground Water Resources and International Law', in *American University International Law Review*, 19 (2003) 201.

Elmusa, S.S. (1995), 'Dividing Common Water Resources According to International Water Law: The case of the Palestinian-Israeli waters', in *Natural Resources Journal*, 35 (1995) 223.

Expert Group on Environmental Law of the World Commission on Environment and Development (1987), *Environmental Protection and Sustainable Development: Legal principles and recommendations*, Nijhoff: Dordrecht.

Farmer, A.M. (2001), 'The EC Water Framework Directive: An introduction', in *Water Law*, 12 (2001) 1, 40-46.

Faruqui, N.I., A.K. Biswas and M.J. Bino (eds) (2001), *Water Management in Islam*, United Nations University Press: Tokyo.

Fathallah, R.M. (1996), 'Water Disputes in the Middle East: An international law analysis of the Israel-Jordan Peace Accord', in *Journal of Land Use and Environmental Law*, 12 (1996) 119.

Faure, M., J. Gupta and A. Nentjes (eds) (2003), *Climate Change and the Kyoto Protocol: The role of institutions and instruments to control global change*, Edward Elgar: Cheltenham.

Fitzmaurice, M.A. (2001), 'International Protection of the Environment', in Hague Academy of International Law, *Collected Courses*, 293 (2001).

Fitzmaurice, M. and O. Elias (2004), *Watercourse Co-operation in Northern Europe – A model for the future*, T.M.C. Asser Press: The Hague.

Fox, H. (ed.) (1992), *International Economic Law and Developing States: An introduction*, London.

Franck, T.M. (1995), *Fairness in International Law and Institutions*, Clarendon Press: Oxford.

Frant, N. (2003), 'Developments in Transboundary Water', in *Colorado Journal of International Environmental Law and Policy yearbook 2002*, 91-99.

Freestone, D. (1999), 'The Challenge of Implementation: Some concluding notes', in Boyle and Freestone (1999), 359-364.

French, D.A. (2002), 'The Role of the State and International Organizations in Reconciling Sustainable Development and Globalization', in *International Environmental Agreements: Politics, Law and Economics*, 2 (2002) 2, 135-150.

Fuentes, X. (1999), 'Sustainable Development and the Equitable Utilization of International Watercourses', in *The British Yearbook of International Law 1998*, (1999), Clarendon Press: Oxford, 119-200.

Fuentes, X. (2002), 'International Law-Making in the Field of Sustainable Development: The unequal competition between development and the environment', in *International Environmental Agreements: Politics, Law and Economics*, 2 (2002) 2, 109-133.

Gandhi, M. (1992), 'Right to Development as a Right to Equal Resources', in Chowdhury, Denters and De Waart (eds) (1992), 139-143.

Geon, B.S. (1997), 'A Right to Ice?: The application of international and national water laws to the acquisition of iceberg rights', in *Michigan Journal of International Law*, 19 (1997) 277.

Gillies, D. (1999), *A Guide to EC Environmental Law*, Earthscan Publications Ltd: London.

Giordano, M.A. and A.T. Wolf (2001), 'Incorporating Equity into International Water Agreements', in *Social Justice Research* 14 (2001) 355.

Girouard, R.J. (2003), 'Water Export Restrictions: A case study of WTO dispute settlement', in *The Georgetown International Environmental Law Review*, 15 (2003) 2, 247-289.

Ginther, K., H.M.G. Denters and P.J.I.M. de Waart (eds) (1995), *Sustainable Development and Good Governance*, Martinus Nijhoff: Dordrecht.

Giorgetta, S. (2002), 'The Right to a Healthy Environment, Human Rigths and Sustainable Development', in *International Environmental Agreements: Politics, Law and Economics*, 2 (2002) 2, 171-192.

Glazewski, J. (1996), 'Environmental Rights and the New South African Constitution', in Boyle and Anderson (eds) (1996), 177-197.

Gleick, P.H. (1998), *The World's Water 1998-1999: The biennial report on freshwater resources*, Island Press: Washington.

Gleick, P.H. (2000), *The World's Water 2000-2001: The biennial report on freshwater resources*, Island Press: Washington.

Gleick, P.H. (2002), 'Water Management: Promoting the Soft Path', in *Nature*, 418 (2002) 373.

Gleick, P.H., W.C.G. Burns, E.L. Chalecki, M. Cohen, K.K. Cushing, A. Mann, R. Reyes, G.H. Wolff, and A.K. Wong (2002), *The World's Water 2002-2003: The biennial report on freshwater resources*, Island Press: Washington.

Gleick, P.H., G. Wolff, E.L. Chalecki and R. Reyes (2002), *The New Economy of Water: The risks and benefits of globalization and privatization of fresh water*, Pacific Institute for Studies in Development, Environment, and Security: Oakland.

Glennon, R.J. (2003), *Water Follies: Groundwater pumping and the fate of America's fresh waters*, Island Press: Washington, D.C.

Global Water Partnership (2000), *Towards Water Security: A framework for action*, GWP: Stockholm.

Godana, B.A. (1985), *Africa's Shared Water Resources: Legal and institutional aspects of the Nile, Niger and Senegal River Systems*, Pinter: London.

Goodman, E.J. (2000), 'Indian Tribal Sovereignty and Water Resources: Watersheds, ecosystems and tribal co-management', in *Journal of Land, Resources, and Environmental Law*, 20 (2000) 185.

Green, C. (2000), 'If Only Life Were That Simple; Optimism and pessimism in economics', in *Physics and Chemistry of the Earth (B)*, 25 (2000) 3, 205-212.

Green Cross International (2000), National Sovereignty and International Watercourses, Green Cross International: Conches-Geneva.

Grünfeld, F. (1994), 'Recht op Voedsel als Voorbeeld van Militaire Implementatie van een Mensenrecht', in Maastricht Centrum voor de Rechten van de Mens (1994), 74-85.

Gupta, J. (1996), *WaterLaw: An introduction to water law, policies and institutions*, IHE: Delft.

Gupta, J. (1997), *The Climate Change Convention and Developing Countries: From conflict to consensus?*, Kluwer Academic: Dordrecht.

Gupta, J. (2001), *Our Simmering Planet: What to do about global warming?*, Zed Books: London.

Gupta, J. (2002), 'Global Sustainable Development Governance: Institutional challenges from a theoretical perspective', in *International Environmental Agreements: Politics, Law and Economics*, 2 (2002) 4, 361-388.

Gupta, J. (2003), 'The Role of Non-State Actors in International Environmental Affairs', in *Zeitschrift für ausländisches öffentliches Recht und Völkerrecht*, 63 (2003) 2, pp 459-486.

Gupta, J. (2004), *(Inter)national Water Law and Governance: Paradigm lost or gained?*, Inaugural Address, UNESCO-IHE: Delft.

Gupta, J. and M. Hisschemöller (1997), 'Issue-Linkages: A global strategy towards sustainable development', in *International Environmental Affairs*, 9 (1997) 4, 289-308.

Haan, E.J. de (1997), 'Balancing Free Trade in Water and the Protection of Water Resources in GATT', in Brans, De Haan, Nollkaemper and Rinzema (eds) (1997), 245-259.

Haas, P.M. (2002), 'UN Conferences and Constructivist Governance of the Environment', in Global Governance, 8 (2002) 1, 73-91.

Hancher, L. (1997), 'Chapter 17 Privatization of Drinking Water in Europe', in Brans, De Haan, Nollkaemper and Rinzema (eds) (1997), 277-289.

Hedger, M., B. Natarajan, J. Turkson and D. Wallace (2000), 'Enabling Environments for Technology Transfer', in Inter-governmental Panel on Climate Change (2000), *Special Report on Technology Transfer*, Cambridge University Press: Cambridge, 105-141.

Hey, E. (1995), 'Sustainable Use of Shared Water Resources: The need for a paradigmatic shift in international watercourses law', in G.H. Blake *et al.* (eds), *The Peaceful Management of Transboundary Resources*, Graham and Trotman: London.

Hey, E. (1998), 'The Watercourses Convention: To what extent does it provide a basis for regulating uses of international watercourses?', in *RECIEL*, 7 (1998) 3, 291-300.

Hey, E. (2003), *Teaching International Law: State-consent as consent to a process of normative development and ensuing problems*, Kluwer Law International: The Hague.

Hey, E. and A. Nollkaemper (1992), 'The Second International Water Tribunal', in *Environmental Policy Law*, 22 (1992), 82-87.

Heyns, P. (2002), 'The Interbasin Transfer of Water between SADC Countries: A developmental challenge for the future', in Turton and Henwood (eds) (2000), 157-176.

Hildering, P.A. (2002), 'Book Reviews: Tanzi/Arcari, The United Nations Convention on the Law of International Watercourses: A Framework for Sharing, and McCaffrey, The Law of International Watercourses: Non-navigational Uses', *NILR* (2002) 3, 422-427.

Hildering, P.A. (2003), 'Waken over Water via het Internationale Recht', in: *VN Forum* 2003-3, Special over water, pp. 5-13.

Hildering, P.A. (2004), 'The Right of Access to Freshwater Resources', in Schrijver and Weiss (eds) (2004), 405-429.

Hildering, P.A. (forthcoming), 'Water as an Economic Good', in Hague Academy of International Law (forthcoming), papers of the 2001 Programme 'Water Resources and International Law' of the Centre for Studies and Research in International Law and International Relation.

Hirji, R. and D. Grey (1998), 'Managing International Waters in Africa: Process and Progress', in Salman and Boisson de Chazournes (eds) (1998), 77-99.

His Royal Highness The Prince of Orange (2002), No Water No Future: A water focus for Johannesburg, Contribution of HRH the Prince of Orange to the Panel of the UN Secretary General in preparation for the Johannesburg Summit, Final Version: August 2002.

His Royal Highness The Prince of Orange and F.R. Rijsberman (2000), 'Summary Report of the Second World Water Forum: From vision to action', in World Water Council (2000), 13-22.

Hoekstra, A.J. (ed.) (2003), *Virtual Water Trade: Proceedings of the International Expert Meeting on Virtual Water Trade*, Value of Water Research Report Series No.12, IHE: Delft.

Hohmann, H. (1992), 'Environmental Implications of the Principle of Sustainable Development and their Realization in International Law', in Chowdhury, Denters and De Waart (eds) (1992), 273-285.

Holland, M.M., E.R. Blood and L.R. Shaffer (2003), *Achieving Sustainable Freshwater Systems: A web of connections*, Island Press: Washington, D.C.

Homer-Dixon, T.F., J.H. Boutwell and G.W. Rathjens (1993), 'Environmental Change and Violent Conflict: Growing scarcities of renewable resources can contribute to social instability and civil strife', in *Scientific American*, February 1993, 16-23.

Hunter, D., J. Salzman and D. Zaelke (2002) (2nd edition), *International Environmental Law and Policy*, Foundation Press: New York.

International Bureau of the Permanent Court of Arbitration (2003), *Resolution of International Water Disputes Water Conflicts*, Kluwer Law International: The Hague.

International Law Association Committee on International Human Rights Law and Practice (2004), *Final Report*, Berlin Conference (2004).

International Law Association Committee on International Law on Sustainable Development (2004), *First Report*, Berlin Conference (2004).

International Law Association Committee on Legal Aspects of Sustainable Development (2000), *Fourth Report*, London Conference (2000).

International Law Association Committee on Legal Aspects of Sustainable Development (2002), *Fifth and Final Report: Searching for the contours of international law in the field of sustainable development*, New Delhi Conference 2002.

International Law Association Committee on Water Resources Law (2000), *Second Report*, London Conference (2000).

International Law Association Committee on Water Resources Law (2004), *Fourth Report*, Berlin Conference (2004).

IUCN – The World Conservation Union (2004) (3rd edition), *Draft International Covenant on Environment and Development*, Commission on Environmental Law of IUCN – The World Conservation Union in co-operation with the International Council of Environmental Law, Environmental Policy and Law Paper No. 31 Rev. 2.

IUCN, UNEP and WWF (1991), *Caring for the Earth: A strategy for sustainable living*, IUCN, UNEP and WWF: Gland.

Kannler, K.A. (2002), 'The Struggle among the States, the Federal Government, and Federally Recognized Indian Tribes to Establish Water Quality Standards for Waters Located on Reservations', in *The Georgetown International Environmental Law Review*, 15 (2002) 1, 53-77.

Karr, J.R. (1993), 'Protecting Ecological Integrity: An urgent societal goal, in *The Yale Journal of International Law*, 18 (1993) 1, 297-306.

Kaya, I. (2003), *Equitable Utilization: The law of the non-navigational uses of international watercourses*, Ashgate: Hampshire.

Kenig-Witkowska, M.M. (1988), 'The UN Declaration on the Right to Development in the Light of its Travaux Preparatoires', in De Waart, Peters and Denters (1988), 381-388.

Kentin, E. (2001), 'Domestic Environmental Regulation in International Investment Dispute Settlement', in *Griffin's View on International and Comparative Law*, 2 (2001) 2, 81-90.

Klabbers, J. (2002), *An Introduction to International Institutional Law*, Cambridge University Press: Cambridge.

Koudstaal, R., F. Rijsberman and H. Savenije (1992), 'Water and Sustainable Development', in *Natural Resources Forum*, 16 (1992) 4, 277-290.

Krishna, R. (1998), 'The Evolution and Context of the Bank Policy for Projects on International Waterways', in Salman and Boisson de Chazournes (eds) (1998), 31-43.

Kroes, M. (1997), 'The Protection of International Watercourses as Sources of Fresh Water in the Interest of Future Generations', in Brans, De Haan, Nollkaemper and Rinzema (1997), 80-99.

Kwiatkowska, B., H. Dotinga, E.J. Molenaar, A.G. Oude Elferink and A.H.A. Soons (eds) (2002), *International Organizations and the Law of the Sea 2000: Documentary Yearbook*, Martinus Nijhoff: The Hague.

Lammers, J.G. (1984), *Pollution of International Watercourses: a Search for Substantive Rules and Principles of Law*, Martinus Nijhoff: The Hague.

Lammers, J.G. (1998), 'The Gabčíkovo-Nagymaros Case Seen in Particular From the Perspective of the Law of International Watercourses and the Protection of the Environment', in *Leiden Journal of International Law*, 11 (1998) 2, 287-320.

LeRoy, P. (1995), 'Troubled Waters: Population and water scarcity', in *Colorado Journal of International Environmental Law and Policy*, 6 (1995) 299.

Li, Y. (1994), *Transfer of Technology for Deep Sea-Bed Mining: The 1982 Law of the Sea Convention and beyond*, Martinus Nijhoff: Dordrecht.

Lien, R.A. (1998), 'Still Thirsting: Prospects for a Multilateral Treaty on the Euphrates and Tigris rivers following the adoption of the United Nations Convention on International Watercourses', in *Boston University International Law Journal*, 16 (1998) 273.

Lipper, J. (1967), *Equitable Utilisation in The Law of International Drainage Basins*, Oceana Publications: New York.

Little, S.P. (1996), 'Canada's Capacity to Control the Flow: Water export and the North American Free Trade Agreement', in *Pace International Law Review*, Vol. 8 (1996), 127-159.

Lopez, M. (1997), 'Border Tensions and the Need for Water: An application of equitable principles to determine water allocation from the Rio Grande to the United States and Mexico', in *Georgetown International Environmental Law Review*, 9 (1997) 489.

Lowe, V. (1999), 'Sustainable Development and Unsustainable Arguments', in Boyle and Freestone (eds) (1999), 19-37.

M'Baye, K. (1972), 'Le droit au développement comme un droit de l'homme', in *Revue des Droits de l'Homme*, V (1972) 2-3.

Maastricht Centrum voor de Rechten van de Mens (1994), *De Toenemende Betekenis van Economische, Sociale en Culturele Mensenrechten*, Stichting NJCM-Boekerij: Leiden

Magat, W.A., J. Huber and W.K. Viscusi (2000), *An Iterative Choice Approach to Valuing Clean Lakes, Rivers, and Streams*, Harvard John M. Olin Discussion Paper Series Paper No. 295.

216

Mariño, M. and K.E. Kemper (eds) (1999), *Institutional Frameworks in Successful Water Markets: Brazil, Spain, and Colorado, USA*, World Bank Technical Paper No. 427.

Matsui, Y. (2002), 'Some Aspects of the Principle of "Common but Differentiated Responsibilities"', in *International Environmental Agreements: Politics, Law and Economics*, 2 (2002) 2, 151-170.

Matsui, Y. (2004), 'The Principle of "Common but Differentiated Responsibilities"', in Schrijver and Weiss (eds) (2004), 73-96.

McCaffrey, S.C. (1992), 'A Human Right to Water: Domestic and International Implications', in *Georgetown International Environmental Law Review*, 5 (1992) 1.

McCaffrey, S.C. (1997), 'Water Scarcity: Institutional and legal responses', in Brans, De Haan, Nollkaemper and Rinzema (eds) (1997), 43-58.

McCaffrey, S.C. (1998), 'The UN Convention on the Law of Non-Navigational Uses of International Watercourses: Prospects and pitfalls', in Salman and Boisson de Chazournes (eds) (1998), 17-28.

McCaffrey, S.C. (2001), *The Law of International Watercourses: Nonnavigational uses*, Oxford University Press: Oxford.

McGee, S. (2002), 'Proposals for Ballast Water Regulation: Biosecurity in an insecure world', in *Colorado Journal of International Environmental Law and Policy Yearbook 2001*, 141-159.

Molen, I. van der (2001), *Rains, Droughts and Dreams of Prosperity: Resourceful strategies in irrigation management and beyond – The Sri Lankan case*, Dissertatie Universiteit Twente: Enschede.

Murty, M.N., A.J. James and S. Misra (1999), *Economics of Water Pollution: The Indian experience*, Oxford University Press: New Delhi.

Nayak, R.K. (1992), 'Evolving Right to Development as a Principle of Human Rights Law', in Chowdhury, Denters and De Waart (eds) (1992), 145-154.

Nelissen, F.A. (2002), *Van Stockholm, via Rio naar Johannesburg: Enige volkenrechtelijke beschouwingen over het beginsel van goed nabuurschap*, Inaugural Address, Rijksuniversiteit Groningen: Groningen.

Nollkaemper, A. (1993), *The Legal Regime for Transboundary Water Pollution: Between discretion and constraint*, Martinus Nijhoff: Dordrecht.

Nollkaemper, A. (1996), 'The Contribution of the International Law Commission to International Water Law: Does it reverse the flight from substance?', in *Netherlands Yearbook of International Law*, XXVII (1996), 39-73.

Okaru-Bisant, V. (1998), 'Institutional and Legal Frameworks for Preventing and Resolving Disputes Concerning the Development and Management of Africa's Shared River Basins', in *Colorado Journal of International Environmental Law and Policy*, 9 (1998) 331.

Pearsall, J. (ed.) (1999) (10th edition), *The Concise Oxford Dictionary*, Oxford University Press Inc.: New York.

Perrez, F.X. (2000), *Cooperative sovereignty: From independence to inter-dependence in the structure of international environmental law*, Kluwer Law International: The Hague.

Pescatore, P., W.J. Davey and A.F. Lowenfeld (eds) (1995), *Handbook of WTO/GATT Dispute Settlement*, Volume I, Transnational Juris Publications, Irvington-on-Hudson, N.Y.

Petrella, R. (2001), *The Water Manifesto: Arguments for a World Water Contract*, Zed Books: London.

Postel, S.L. (1996), *Dividing the Waters: Food security, ecosystem health, and the new politics of scarcity*, Worldwatch Paper 132.

Postel, S. and B. Richter (2003), *Rivers for Life: Managing water for people and nature*, Island Press: Washington, D.C.

Picolotti, R. and J.D. Taillant (eds) (2003), *Linking Human Rights and the Environment*, University of Arizona Press: Tucson.

Rieu-Clarke, A.S. (2004), 'Sustainable Use and the EU Water Framework Directive: From principle to practice?', in Schrijver and Weiss (eds) (2004), 557-574.

Rockström, J., C. Figuères and C. Tortajada (2003), *Rethinking Water Management: Innovative approaches to contemporary issues*, Earthscan: London.

Sachs, A. (1995), *Eco-Justice: Linking human rights and the environment*, Worldwatch Paper 127.

Salman, S.M.A. (1997), *The Legal Framework for Water Users' Associations: A comparative study*, World Bank Technical Paper No. 360.

Salman, S.M.A. (1999), *Groundwater: Legal and Policy Perspectives, Proceedings of a World Bank Seminar*, World Bank Technical Paper No. 456.

Salman, S.M.A. (2001a), 'Legal Regime for Use and Protection of International Watercourses in the Southern African Region: Evolution and context', in *Natural Resources Journal*, 41 (2001) 981.

Salman, S.M.A. (2001b), 'Dams, International Rivers, and Riparian States: An analysis of the recommendations of the World Commission on Dams', in *American University International Law Review*, 16 (2001) 1477.

Salman, S.M.A. and L. Boisson de Chazournes (eds) (1998), *International Watercourses: Enhancing cooperation and managing conflict, Proceedings of a World Bank Seminar*, World Bank Technical Paper No. 414.

Salman, S.M.A. and S. McInerney-Lankford (2004), *The Human Right to Water: Legal and policy dimensions*, World Bank: Washington, D.C.

Salman, S.M.A. and K. Uprety (2002), Conflict and Cooperation on South Asia's International Rivers: A legal perspective, Kluwer International Law: The Hague.

Sánchez, R.A. (1997), 'Chapter 16 Water Conflicts Between Mexico and the United States: Towards a transboundary regional water market?', in Brans, De Haan, Nollkaemper and Rinzema (eds) (1997), 260-276.

Sands, P. (ed.) (1993), *Greening International Law*, Earthscan: London.

Sands, P. (1995), 'International Law in the Field of Sustainable Development', in *British Year Book of International Law 1994*, 65 (1995), 303-381.

Sands, P. (1998), 'Watercourses, Environment and the International Court of Justice: The Gabcikovo-Nagymaros case', in Salman and Boisson de Chazournes (eds) (1998), 103-125.

Sands, P. (2003) (2nd edition), *Principles of International Environmental Law*, Cambridge University Press: New York.

Saunders, J.O. (1994), 'NAFTA and the North American Agreement on Environmental Cooperation: A new model for international collaboration on trade and environment', in *Colorado Journal of International Environmental Law and Policy*, 5 (1994) 273.

Savenije, H.H.G. (2002), 'Why Water is not an Ordinary Economic Good, or Why the Girls is Special', in *Physics and Chemistry of the Earth*, 27 (2002) 741-744.

Savenije, H.H.G. and P. van der Zaag (2000a), 'The Maseru Conference', in *Water Policy*, 2 (2000) 1-7.

Savenije, H.H.G. and P. van der Zaag (2000b), 'Conceptual Framework for the Management of Shared River Basins; With special reference to the SADC and EU', in *Water Policy*, 2 (2000) 9-45.

Savenije, H.H.G. and P. van der Zaag (2002), 'Water as an Economic Good and Demand Management: Paradigms with pitfalls', in *Water International*, 27 (2002) 1, 98-104.

Scanlan, K.P. (1996), 'The International Law Commission's First Ten Draft Articles on the Law of the Non-Navigational Uses of International Watercourses: Do they adequately address all the major issues of water usage in the Middle East?', in *Fordham International Law Journal*, 19 (1996) 2180.

Scanlon, J., A. Cassar and N. Nemes (2004), *Water as a Human Right?*, IUCN Environmental Policy and Law Paper No. 51.

Schachter, O. (1977), *Sharing the World's Resources*, Columbia University Press: New York, N.Y.

Schachter, O. (1991), *International Law in Theory and Practice*, Martinus Nijhoff: Dordrecht.

Schachtschneider, K. (2002), 'Water Demand Management and Tourism in Arid Countries: Lessons from Namibia', in Turton and Henwood (eds) (2002), 205-215.

Schermers, H.G. and N.M. Blokker (2003) (4th edition), *International Institutional Law: Unity within diversity*, Martinus Nijhoff: Leiden.

Schorer, K.F.H. (ed.) (1992), *De Waterschapswet Beschouwd*, Unie van Waterschappen: 's-Gravenhage.

Schrijver, N.J. (1997), *Sovereignty over Natural Resources: Balancing rights and duties*, Cambridge University Press: Cambridge.

Schrijver, N.J. (2000), 'The Changing Nature of State Sovereignty', in *The British Year Book of International Law 1999*, Oxford University Press: Oxford , 65-98.

Schrijver, N.J. (2001), *On the Eve of Rio+10: Development – the neglected dimension in the international law of sustainable development*, ISS: The Hague.

Schrijver, N.J. and F. Weiss (eds) (2004), *International Law and Sustainable Development. Principles and practice*, Martinus Nijhoff: Leiden.

Schuttelar, M., V. Ozbilen, T. Ikeda, M. Hua, F. Guerquin and T. Ahmed (2003), *World Water Actions: Making water flow for all*, Earthscan: London.

Schwabach, A. (1998), 'The United Nations Convention on the Law of Non-navigational Uses of International Watercourses, Customary International Law, and the Interests of Developing Upper Riparians', in *Texas International Law Journal*, 33 (1998) 257, 257-279.

Schwabach, A. (2000), 'From Schweizerhalle to Baia Mare: The continuing failure of international law to protect Europe's rivers', in *Virginia Environmental Law Journal*, 19 (2000) 431.

Sen, A. (1999), *Development as Freedom*, Oxford University Press: Oxford.

Shaw, M.N. (1997) (4th edition), *International Law*, Cambridge University Press: Cambridge.

Shelton, D. (ed.) (2000), *Commitment and Compliance: The role of non-binding norms in the international legal system*, Oxford University Press: Oxford.

Smets, H. (2000), 'The Right to Water as a Human Right', in Environmental Policy and Law, 30 (2000) 5, 248.

Solanes, M. and F. Gonzalez-Villarreal (1999), *The Dublin Principles for Water as Reflected in a Comparative Assessment of Institutional and Legal Arrangements for Integrated Water Resources Management*, Technical Advisory Committee, Global Water Partnership: Chile.

Stec, S. (1999), 'Do Two Wrongs Make A Right? Adjudicating sustainable development in the Danube Dam Case', in *Golden Gate University Law Review*, 29 (1999) 317.

Steiner, H.J. and P. Alston (2000) (2nd edition), *International Human Rights in Context*, Oxford University Press: Oxford.

Tanzi, A. (1998), 'The UN Convention on International Watercourses as a Framework for the Avoidance and Settlement of Waterlaw Disputes', in *Leiden Journal of International Law*, 11 (1998) 3, 441-472.

Tanzi, A. and M. Arcari (2001), *The United Nations Convention on the Law of International Watercourses*, Kluwer Law International: London.

Tarlock, A.D. (1997), 'The Missouri River: The paradox of conflict without scarcity', in *Great Plains Natural Resources Journal*, 2 (1997) 1.

Tarlock, A.D. (2000), 'How well can International Water Allocation Regimes adapt to Global Climate Change?', in *Journal of Land Use and Environmental Law*, 15 (2000) 423.

Teclaff, L.A. (1967), *The River Basin in History and Law*, Martinus Nijhoff: The Hague.

Teclaff, L.A. (1985), *Water Law in Historical Perspective*, William S. Hein Company: Buffalo.

Teclaff, L.A. (1994), 'Restoring River and Lake Basin Ecosystems', in *Natural Resources Journal*, 34 (1994) 905.

Teclaff, L.A. (1996), 'Evolution of the River Basin Concept in National and International Water Law', in *Natural Resources Journal*, 36 (1996) 359.

Toebes, B.C.A. (1999), *The Right to Health as a Human Right in International law*, Intersentia: Antwerpen.

Turton, A. and R. Henwood (eds) (2002), *Hydropolitics in the Developing World: A Southern African perspective*, African Water Issues Research Unit (AWIRU): Pretoria.

UNDP (1997), *Governance for Sustainable Human Development: A UNDP policy document*, UNDP: New York.

United Nations Human Settlement Programme (UN-HABITAT) (2003), *Water and Sanitation in the World's Cities: Local action for global goals*, Earthscan: London.

Utton, A. (1982), 'The Development of International Ground Water Law', *Natural Resources Journal*, 22 (1982) 95.

Vass, P. (2002), 'Competition and Restructuring in the UK Water Industry', in *Journal of Network Industries*, 3 (2002) 1, 77-98.

Vellinga, P. and M. van Drunen (eds) (1999) (3rd edition), *The Environment: A multidisciplinary concern*, Institute for Environmental Studies: Amsterdam.

Verbruggen, H. (1999), 'Environment, International Trade and Development', in J.C.J.M. van den Bergh (ed.) (1999), *Handbook of Environmental and Resource Economics*, Edward Elgar: Cheltenham, pp. 449-460.

Vethaak, A.D., G.B.J. Rijs, S.M. Schrap, H. Ruiter, A. Gerritsen, J. Lahr (2002), *Estrogens and Xeno-estrogens in the Aquatic Environment of the Netherlands: Occurrence, potency and biological effects*, RIZA/RIKZ report No. 2002.001.

Villiers, M. de (2001), *Water: The fate of our most precious resource*, Mariner Books: Boston.

Vinogradov, S. (1996), 'Transboundary Water Resources in the Former Soviet Union: Between conflict and cooperation', in *Natural Resources Journal*, 36 (1996) 393.

Waart, P.J.I.M. de (1988), 'State Rights and Human Rights as Two Sides of One Principle of International Law: The right to development', in De Waart, Peters and Denters (1988), 371-380.

Waart, P.J.I.M. de (1997), 'Securing Access to Safe Drinking Water through Trade and International Migration', in Brans, De Haan, Nollkaemper and Rinzema (eds) (1997), 100-118.

Waart, P.J.I.M. de, P. Peters and H.M.G. Denters (eds) (1988), *International Law and Development*, Martinus Nijhoff: Dordrecht.

Weiss, F., H.M.G. Denters and P.J.I.M. de Waart (eds) (1998), *International Economic Law with a Human Face*, Kluwer Law International: The Hague.

Westendorp, I. (1994), 'Internationale Implementatie van het Recht op Behoorlijke Huisvesting', in Maastricht Centrum voor de Rechten van de Mens (1994), 100-115.

Wester, P. and J. Warner (2002), 'River Basin Management reconsidered', in Turton and Henwood (2002), 61-71.

Wiebe, K. (2001), 'The Nile River: Potential for conflict and cooperation in the face of water degradation', in *Natural Resources Journal*, 41 (2001) 731.

Wolf, A.T., J.A. Natharius, J.J. Danielson, B.S. Ward, J.K. Pender (1999), 'International River Basins of the World', in International Journal of Water Resources Development, 15 (1999) 4, 387-427.

Woodley, S., J. Kay and G. Francis (eds) (1993), *Ecological Integrity and the Management of Ecosystems*, St. Lucie Press: Delray Beach.

World Bank (2003), *Environment Matters*, Annual Review 2003, The World Bank Group: Washington, D.C.

World Bank (2003b), *Water Resources Sector Strategy: Strategic directions for World Bank Engagement*, The World Bank Group: Washington, D.C.

World Commission for Water in the 21st Century (2000), *A Water Secure World: Vision for water, life, and the environment*, Thanet Press.

World Commission on Dams (2000), *Dams and Development: A new framework for decision-making*, The Report of the World Commission on Dams, Earthscan Publishing: London.

World Commission on Environment and Development (1987), *Our Common Future*, Oxford University Press: Oxford.

World Commission on the Social Dimension of Globalization (2004), *A Fair Globalization: Creating opportunities for all*, ILO Publications: Geneva.

World Health Organization (2003), *Right to Water*, Health and human rights publication series No. 3.

World Water Assessment Programme (2003), *Water for People, Water for Life: The United Nations World Water Development Report*, UNESCO Publishing: Paris.

World Water Council (2000), *Final Report Second World Water Forum & Ministerial Conference*, 17-22 March 2000, The Hague.

Wouters, P.K. (ed.) (1997), *International Water Law: Selected writings of Professor Charles B. Bourne*, Kluwer Law International: The Hague.

Wouters, P.K. and A.S. Rieu-Clarke (2001), *The Role of International Water Law in Promoting Sustainable Development*, Bonn – Sustainable Development Paper, November 2001.

Zaag, P. van der and H. Savenije (1999), 'The Management of International Waters in EU and SADC Compared', in *Physics and Chemistry of the Earth (B)*, 24 (1999) 6, 579-589.

Zaag, P. van der and H. Savenije (2000), 'Towards Improved Management of Shared River Basins: Lessons from the Maseru Conference', in *Water Policy*, 2 (2000) 47-63.

Zaag, P. van der and Á.C. Vaz (2003), 'Sharing the Incomati Waters: Co-operation and competition in the balance', in *Water Policy*, 5 (2003) 349-368.

Zwaan, B. van der and A. Petersen (eds) (2003), *Sharing the Planet: Population – consumption – species, science and ethics for a sustainable and equitable world*, Eburon: Delft.

Index